D1451763

THE SOCIAL ORIGINS
OF LANGUAGE

THE SOCIAL ORIGINS OF LANGUAGE

ROBERT M. SEYFARTH
AND
DOROTHY L. CHENEY

Edited and Introduced by Michael L. Platt

PRINCETON UNIVERSITY PRESS
PRINCETON AND OXFORD

Published by Princeton University Press, 41 William Street,
Princeton, New Jersey 08540

In the United Kingdom: Princeton University Press,
6 Oxford Street, Woodstock, Oxfordshire OX20 1TR

Cover photo by Dorothy L. Cheney

press.princeton.edu

ISBN 978-0-691-17723-6

Library of Congress Control Number 2017952280

British Library Cataloging-in-Publication Data is available

This book has been composed in Sabon Next

Printed on acid-free paper ∞

Printed in the United States of America

1 3 5 7 9 10 8 6 4 2

CONTENTS

The Contributors   vii

INTRODUCTION   1
  Michael L. Platt

PART 1

THE Social Origins of Language   9
  Robert M. Seyfarth and Dorothy L. Cheney

PART 2

1. Linguistics and Pragmatics   37
  John McWhorter

2. Where Is Continuity Likely to Be Found?   46
  Ljiljana Progovac

3. Fluency Effects in Human Language   62
  Jennifer E. Arnold

4. Relational Knowledge and the Origins of Language   79
  Benjamin Wilson and Christopher I. Petkov

5. Primates, Cephalopods, and the Evolution of
  Communication   102
  Peter Godfrey-Smith

PART 3

CONCLUSION   123
  Robert M. Seyfarth and Dorothy L. Cheney

Notes   131

References   135

Index   163

The Contributors

Dorothy Cheney is Professor of Biology and **Robert Seyfarth** is Professor of Psychology at the University of Pennsylvania. They have spent over 40 years studying the social behavior, cognition, and communication of monkeys in their natural habitat.

Michael Platt is Professor of Neuroscience, Psychology, and Marketing at the University of Pennsylvania, and Director of the Wharton Neuroscience Initiative. Michael's work focuses on the biological mechanisms underlying decision making and social behavior, development of improved therapies for treating impairments in these functions, and translation of these discoveries into applications that can improve business, education, and public policy.

John McWhorter specializes in language change and contact and teaches linguistics at Columbia University. He has written *The Power of Babel*, *Our Magnificent Bastard Tongue*, *Words on the Move*, and other books on language, including the academic books *Language Interrupted* and *Defining Creole*.

Ljiljana Progovac is a linguist, with research interests in syntax, Slavic syntax, and evolution of syntax, and is currently professor and director of the Linguistics Program at Wayne State University in Detroit. Among other publications, she is the author of *Evolutionary Syntax* (Oxford University Press, 2015) and *Negative and Positive Polarity* (Cambridge University Press, 1994).

Jennifer Arnold is a professor at the University of North Carolina at Chapel Hill, and received her Ph.D. from Stanford University. She studies the cognitive mechanisms of language use in both adults and children, examining how language is used and understood in social and discourse contexts.

Benjamin Wilson is a Sir Henry Wellcome Fellow in the Laboratory of Comparative Neuropsychology at the Newcastle University School of Medicine. His research focuses on understanding the neurobiological systems that support language, and the evolution of these brain networks. His work combines a range of behavioral and neuroscientific techniques to investigate the extent to which cognitive abilities underpinning human language might be shared by nonhuman primates, and how far these abilities are supported by evolutionarily conserved networks of brain areas.

Christopher I. Petkov is a professor of neuroscience at the Newcastle University School of Medicine. His research is guided by the notion that information on how the human brain changed during its evolutionary history will be indispensable for understanding how to improve treatments for human brain disorders. His Laboratory of Comparative Neuropsychology uses advanced imaging and neurophysiological methods to study perceptual awareness and cognition, with an emphasis on communication, in particular auditory and multisensory communication.

Peter Godfrey-Smith is Professor of the History and Philosophy of Science at the University of Sydney, Australia. His main research interests are the philosophy of biology and the philosophy of mind, though he also works on pragmatism and various other parts of philosophy. He has written five books, including the widely used textbook *Theory and Reality: An Introduction to the Philosophy of Science* (Chicago, 2003) and *Darwinian Populations and Natural Selection* (Oxford, 2009), which won the 2010 Lakatos Award. His most recent book is *Other Minds: The Octopus, the Sea, and the Deep Origins of Consciousness* (Farrar, Straus and Giroux 2016).

THE SOCIAL ORIGINS
OF LANGUAGE

INTRODUCTION

Michael L. Platt

In the beginning was the Word.
—*Genesis 1:1*

Or perhaps it was the grunt. The origins of human language, arguments from religion notwithstanding, remain hotly debated. From a scientific standpoint, human language must have evolved. As the great biologist Theodosius Dobzhansky said: "nothing in biology makes sense except in the light of evolution" (Dobzhansky 1973). Yet, so far, evolutionary accounts have largely failed to provide a comprehensive explanation for why and how human language could have emerged from the communication systems found in our closest primate cousins. This dilemma reflects the fact that communication in human language arises from the union of semantics—words have referents—and syntax—words can be combined according to a set of rules into phrases and sentences capable of generating countless possible messages. Put simply, there is no single nonhuman animal—primate or otherwise—whose natural communication system possesses both semantics and syntax.

This apparent discontinuity has led some (Berwick and Chomsky 2011; Bolhuis et al. 2014) to propose that human language appeared fully formed within the brain of a single human ancestor, like Venus springing from the head of Zeus, solely to support self-directed thought. Only later, according to this view, after language was passed down to the offspring of this Promethean protohuman, did language become a tool for communication. This solipsistic account, however, ignores emerging evidence for continuity in cognitive functions, like episodic memory (Templer and Hampton 2013), decision-making (Santos and Platt 2014; Pearson, Watson, and Platt 2014; Santos and Rosati 2015), empathy (Chang, Gariépy, and Platt 2013; Brent et al. 2013), theory of mind (Klein, Shepherd, and Platt 2009; Drayton and Santos 2016), creativity and exploration (Hayden, Pearson, and Platt 2011; Pearson, Watson, and Platt 2014), counterfactual thinking (Hayden, Pearson, and Platt 2009), intuitive mathematics (Brannon and Park 2015), self-awareness (Toda and Platt 2015), and conceptual thinking (MacLean, Merritt, and Brannon 2008), and the neural circuits that mediate these functions (Platt and Ghazanfar 2010)—though, to be sure, other discontinuities remain, in particular the ability to refer to the contents of representations (so-called ostensive communication or metarepresentations: Sperber and Wilson 1986; Sperber 2000). Despite growing appreciation for cognitive and neural continuity between humans and other animals, an evolutionary account of human language—in its full-blown, modern form—remains as elusive as ever.

This book attempts to provide a new perspective on this quandary and chart a novel pathway toward its resolution. We contend that any biologically and humanistically plausible answer to the question of the origins of language must reflect the combined wisdom of multiple disciplines, each providing a unique but related perspective. In this brief volume, we provide an open dialogue among experts in animal communication, neurobiology, philosophy, psychology, and linguistics that began with a two-day symposium convened by the Duke Institute for Brain Sciences in 2014, at Duke University in Durham, North Carolina. The symposium and accompanying book orbit a keynote lecture by Robert Seyfarth and a provocative target article coauthored by Seyfarth and his long-time collaborator Dorothy Cheney.

Seyfarth and Cheney are well known for their long-term studies of the behavior of monkeys and baboons in the wild, in which they use audio playback of communication calls to probe how primates think about their worlds. In their much-heralded and popular book *How Monkeys See the World* (1990), Seyfarth and Cheney provided strong evidence that vervet monkeys in Amboseli National Park, Kenya, use communication calls that seem to function much like human words, effectively labeling important objects and events in the environment such as predatory eagles and snakes. Taking a fresh look at their own work on social communication among baboons in the Okavango Delta of Botswana, which was originally described in their book *Baboon Metaphysics* (2007), Seyfarth and Cheney argue that the grunts given by baboons in advance of friendly interactions, and the shrieks given in response to aggression, demonstrate not only a richness and complexity in how these animals think about others in their groups but, more surprisingly, that baboons seem capable of combining a small number of communication calls with the large number of individual relationships within the group to produce a vast number of possible messages about social interactions. Seyfarth and Cheney provocatively suggest that these findings provide evidence that the interaction of primate communication systems with cognitive systems representing social knowledge effectively translate into a rudimentary "language" capable of both semantics and generative grammar. For Seyfarth and Cheney, the key elements of human language emerge from the need to decipher and encode complex social interactions in a large, multilayered group.

This bold hypothesis serves as the jumping-off point for a targeted series of responses by symposium participants from several distinct disciplines. These rejoinders situate Seyfarth and Cheney's hypothesis, and the evidence upon which it is based, within the relevant contexts of linguistics, sociology, philosophy, psychology, and neuroscience. The authors find sometimes surprising consilience in the comparison, and sometimes equally surprising contrasts as well.

For example, John McWhorter, a linguist with broad interests in creole languages, finds great resonance with Seyfarth and Cheney's arguments. McWhorter finds commonality in the pragmatics of language—the ways in which context and emphasis markers add new layers of meaning to an utterance—and the complex layering of structured communication in baboon social communication. He argues against a naively Chomskyan "syntactocentrism" and favors theories of language evolution in which pragmatics and semantics precede formal grammar, a view aligned with Seyfarth and Cheney's. In his view, focusing on the complexity of modern languages with a long history of development may be a red herring. After all, pidgin languages possess minimal grammatical machinery yet efficiently convey precise information via pragmatic markers, consistent with a socially based origin for full-blown language.

By contrast, Ljiljana Progovac, a linguist who specializes in Slavic syntax, flips Seyfarth and Cheney's approach on its head by arguing that rather than look for the antecedents of human language in animal communication, we ought instead to look for elements of animal communication systems in human language. Such "living fossils" as it were, for example, two-word combinations that function as a protosyntax, invite the possibility of continuity in the evolution of human language from primate communication.

Jennifer Arnold, a psychologist who focuses on prosody—the timing, pitch, rhythm, and acoustic properties of speech—sympathizes with this perspective as well. Her research emphasizes the impact of subconscious processing routines that somewhat automatically shade spoken language by altering speech timing, pitch, and rhythm. Such markers can betray informational redundancies or statistical regularities that may be exploited by listeners in conversation. It's easy to imagine that the baboons studied by Cheney and Seyfarth use contextual information attending grunts and shrieks to develop a savvy understanding of their social worlds.

Notably, the two more biologically oriented commentaries—one by Peter Godfrey-Smith, a philosopher of biology, and the other by Benjamin Wilson and Christopher Petkov, both neuroscientists—find some agreement with Seyfarth and Cheney but identify significant challenges to their proposal as well. Both chapters make the clear distinction between sender and receiver, and that what is unique about human language is that syntax allows for generative creation of an infinite number of messages by the sender and their interpretation by the receiver. The generative nature of baboon social communication appears to reside entirely within the receiver. Wilson and Petkov compare the impressive sensitivity of baboons to social order as expressed through sequences of calls with Artificial Grammar studies showing monkeys and other animals are sensitive to ordered sequences of arbitrary stimuli, and suggest that in fact social communication may be the prerequisite for the evolution of human language. They also sketch an outline of the neural circuits involved in sequence learning and production, and speculate that these circuits may interact with brain regions involved in social information processing when baboons or other animals make inferences about the interactions of others based on sequences of calls they hear.

Godfrey-Smith provides the most provocative challenge to Seyfarth and Cheney's hypothesis by way of comparing the social communication system of baboons with the social communication systems of cephalopods—squid, octopuses, and cuttlefish. In his view, all the sophistication in baboon communication lies within the receiver. When a baboon emits a call, she surely intends to signal something about the environment—response to a threat, approach to a dominant baboon—yet the possible sets of messages are limited. Nevertheless, baboons listening to sequences of calls made by others can draw far more sophisticated conclusions about their social worlds, which Godfrey-Smith describes as a fortuitous consequence of baboon social ecology and the statistical regularities of vocalizations within the group. By contrast, certain species of cephalopods have evolved elaborate, combinatorial patterns of sequential coloration changes on their skin that, apparently, have very little effect on receivers and, instead, appear to be fortuitous byproducts of internal processes. The comparison of baboons and cephalopods highlights the importance of both sender and receiver in communication, and the fact that all elements of human language—semantics, syntax, pragmatics—must be considered in any account of its evolution.

In the final chapter, Seyfarth and Cheney provide a synthesis of the chapters written by the other authors in response to their own target article. Seyfarth and Cheney find common ground with the other authors in the importance of pragmatics, in addition to semantics and syntax, for shaping the meaning of communication signals. Indeed, all authors seem to agree that primate communication systems provide a rich pragmatic system for representing information about the social world. Ultimately, Seyfarth and Cheney contend, the need for our primate ancestors to represent and convey information about social context was the biological foundation upon which much more complex aspects of human language were scaffolded by evolution.

The foregoing overview makes plain that we have much to learn about how we came to be the only animal on earth with true language. The chapters included here provide a thought-provoking set of interrelated lenses through which we might catch a glimpse of how human language evolved. The ideas summoned in this brief, yet powerful, book endorse the hypothesis that we will answer this, and other challenging questions, only through interdisciplinary dialogue and investigation.

PART 1

THE SOCIAL ORIGINS
OF LANGUAGE

ROBERT M. SEYFARTH AND
DOROTHY L. CHENEY

Human language poses a problem for evolutionary theory because of the striking discontinuities between language and the communication of our closest animal relatives, the nonhuman primates. How could language have evolved from the common ancestor of these two very different systems?

The qualitative differences between language and nonhuman primate communication are now well known (see Fitch 2010 for review). All human languages are built up from a large repertoire of learned, modifiable sounds. These sounds are combined into phonemes, which are combined into words, which in turn are combined according to grammatical rules into sentences. In sentences, the meaning of each word derives both from its own, stand-alone meaning and from its function role as a noun, verb, or modifier. Grammatical rules allow a finite number of elements to convey an infinite number of meanings: the meaning of a sentence is more than just the summed meanings of its constituent words. Languages derive their communicative power from being discrete, combinatorial, rule-governed, and open-ended computational systems, like the number system or the use of 1s and 0s in a digital computer (Jackendoff 1994; Pinker 1994).

By contrast, nonhuman primates (prosimians, monkeys, and apes)—and indeed most mammals—have a relatively small number of calls in their vocal repertoire. These calls exhibit only slight modification during development; that is, their acoustic structure appears to be largely under genetic control (see Hammerschmidt and Fischer 2008 for review). Furthermore, while animals can give or withhold calls voluntarily and modify the timing of vocal production (reviewed in Seyfarth and Cheney 2010), different call types are rarely given in rule-governed combinations (but see Ouattara, Lemasson, and Zuberbuhler 2009; Zuberbuhler 2014). When call combinations do occur, there is little evidence that individual calls play functional roles as agents, actions, or patients. As a result, primate vocalizations, when compared to language, appear to convey only limited information (Bickerton 1990; Hurford 2007; Fitch 2010).

Differences between human language and nonhuman primate communication are most apparent in the domain of call production. Continuities are more apparent, however, when one considers the neural and cognitive mechanisms that underlie call perception, and the social function of language and communication in the daily lives of individuals. Here we begin by briefly reviewing the evidence that homologous brain mechanisms in human and nonhuman primates underlie the recognition of individual faces and voices; the multimodal processing of visual and auditory signals; the recognition of objects; and the recognition of call meaning. These results are relevant to any theory of language evolution because they suggest that, for much of their shared evolutionary history, human and nonhuman primates faced similar communicative problems and responded by evolving similar neural mechanisms.

What were these similar communicative problems? In the second part of this essay we compare how language functions in human social interactions with the function of vocalizations in the daily lives of animals, particularly baboons. We argue that, while the two systems of communication are structurally very different, they share many functions. These shared functions help explain the evolution of homologous neural mechanisms.

To understand the function of primate vocalizations, one must understand what primates know about each other. In baboons, for example, this includes knowledge of individual identity, dominance rank, matrilineal kin membership, and the use of different vocalizations in different social circumstances. In chimpanzees, it includes knowledge of other animals' alliance partners. In the third part of this essay we show that selection has favored in baboons—and, by extension, other primates—a system of communication that is discrete, combinatorial, rule-governed, and open-ended. We argue that this system was common to our prelinguistic primate ancestors and that when language later evolved from this common foundation, many of its distinctive features were already in place.

SHARED BRAIN MECHANISMS

An area in the human temporal cortex, the fusiform face area, responds especially strongly to the presentation of faces and appears to be specialized for face recognition (Kanwisher, McDermott, and Chun 1997). A similar area, consisting entirely of face-selective cells, exists in the macaque temporal cortex (Tsao et al. 2003, 2006; Freiwald, Tsao, and Livingston 2009). Humans also have a region in the superior temporal sulcus that is particularly responsive to human voices and appears to play an important role in voice recognition (Van Lancker et al. 1988; Belin et al. 2000; Belin and Zattore 2003). Petkov et al. (2008) document the existence of a similar area in the macaque brain.

When communicating with one another, humans exhibit multisensory integration: bimodal stimuli (voices and concurrent facial expressions) consistently elicit stronger neural activity than would be elicited by either voices or faces alone (e.g., Wright et al. 2003). Human infants are sensitive to the "match" between speech sounds and their corresponding facial expressions, responding more strongly to incongruent than to congruent vocal and visual stimuli (Kuhl and Meltzoff 1984; Patterson and Werker 2003). A variety of studies document similar multisensory integration in monkeys (Ghazanfar and Logothetis 2003; Ghazanfar et al. 2005; Ghazanfar, Chandrasekaran, and Logothetis 2008; Sliwa et al. 2011; Adachi and Hampton 2012).

In both humans and macaques, neurons in the ventral premotor cortex exhibit neural activity both when performing a specific action and when observing another perform the same action (see Ferrari, Bonini, and Fogassi 2009 for review). These "mirror neurons" seem likely to be involved in the development of novel behaviors and may constitute a shared, homologous neural substrate for imitative behavior (Ferrari, Bonini, and Fogassi 2009; de Waal and Ferrari 2010).

We take it for granted that humans can classify words according to either their meaning or their acoustic properties. Judged according to their meaning, *treachery* and *deceit* are alike whereas *treachery* and *lechery* are different; judged according to their acoustic properties, these assessments would be reversed. The "ape language" projects were the first to suggest that, like humans, nonhuman primates can classify communicative signals according to either their physical properties or their meaning (Premack 1976; Savage-Rumbaugh et al. 1980); field experiments using vocalizations are consistent with this view (Cheney and Seyfarth 1990; Zuberbuhler, Cheney, and Seyfarth 1999). In a study of the underlying neural mechanisms, Gifford et al. (2005) found that, as in humans, the ventrolateral prefrontal cortex plays an important role in the classification of conspecific calls with different acoustic properties that either are or are not associated with the same events in the animals' daily lives.

From fMRI studies of human cognition, there is increasing evidence that we respond to object words and to the sight of objects using a distributed perceptual representation based on an object's physical features, the motor movements used to interact with it, and a "semantic representation" based on previously acquired information (Martin 1998:72; Barsalou et al. 2003; Yee, Drucker, and Thompson-Schill 2010). Preliminary evidence supports the view that there exists, in monkeys, "a homologous system . . . for representing object information" (Gil da Costa et al. 2004:17518; see also Cheney and Seyfarth 2007:241–243).

To summarize, human and nonhuman primates share many neurological mechanisms for perceiving, processing, and responding to communicative signals. These shared mechanisms are unlikely to have arisen by accident. Instead, it seems likely that during their long, common evolutionary history (roughly 35–25 million years ago: Steiper, Young, and Sukarna 2004), Old World monkeys, apes, and humans faced similar problems in communication and evolved homologous mechanisms to deal with them. The unique, more recent evolution of language in the human lineage (during the past 5–6 million years: Enard et al. 2002) built upon these shared mechanisms. What were the common communicative problems that gave rise to them? How is language used in human social interactions, and how does its use compare with the function of vocalizations in the social interactions of monkeys and apes?

THE SOCIAL FUNCTION OF LANGUAGE

In his review and analysis of language use, Herbert Clark proposes that language is a form of joint action, used by people to facilitate and coordinate their activities. The individuals involved, moreover, are not "generic speakers and addressees, but real people, with identities, genders, histories, personalities, and names" (1996:xi). As will become clear, the parallels with monkeys could hardly be more striking.

Clark (1996:23–24) offers six propositions that characterize how language functions in the daily lives of humans. We repeat these propositions here because they provide an ideal background against which to compare the social function of language with the social function of vocalizations in the lives of primates.

- *Language is used for social purposes.* People don't just use language, they use it for doing things: gossiping, manipulating, planning, and so on. "Languages as we know them wouldn't exist if it weren't for the social activities" in which they play an instrumental role.
- *Language use is a type of joint action* that requires a minimum of two agents and the coordination of activities.
- *Language use always involves speaker's meaning and addressee's understanding.* "We are not inclined to label actions as language use unless they involve one person meaning something for another person who is in a position to understand what the first person means."
- *The basic setting for language use is face-to-face conversation.* "For most people conversation is the commonest setting of language use, . . . and if conversation is basic, then other settings are derivative in one respect or another."
- *Language use often has more than one layer of activity.* While "conversation, at its simplest, has only one layer of action . . . any participant can introduce further layers by telling stories or play-acting at being other people."
- *The study of language use is both a cognitive and a social science.* While "cognitive scientists have tended to study speakers and listeners as [isolated] individuals, . . . social scientists have tended to study language use primarily as a joint activity. [But] if language use truly is a species of joint activity, it cannot

be understood from either perspective alone." It must be both a cognitive and a social science.

Here we use Clark's propositions as a starting point from which to compare the social function of language and the social function of nonhuman primate vocalizations.

THE SOCIAL FUNCTION OF ANIMAL VOCALIZATIONS
Theoretical Background

Animals often compete: over food, a mate, a territory, or some other resource. But rather than escalate immediately to physical fighting, individuals typically engage in nonaggressive communicative displays, like the roaring of red deer (Clutton-Brock et al. 1979), the "jousting" displays of stalk-eyed flies (Wilkinson and Dodson 1997), the croaking of European toads (Davies and Halliday 1978), or the loud "wahoo" calls of male baboons (Kitchen, Cheney, and Seyfarth 2003). Ethologists now have a good understanding of how these displays have evolved—that is, why they are evolutionarily stable. In red deer, for example, roaring is energetically costly, so only males in good physical condition can roar repeatedly, for long durations (Clutton-Brock and Albon 1979). Moreover, the acoustic features of a male's roar are constrained by his body size, so only large males can produce deep-pitched roars (Reby et al. 2005). And larger males are more successful fighters (Clutton-Brock et al. 1979). As a consequence, a male's roaring cannot be faked—because small males and males in poor condition cannot produce low-pitched roars at a high rate—and roaring serves as an honest indicator of size, condition, and competitive ability.

Natural selection has therefore favored listeners who decide whether to escalate or retreat based on their opponent's roars. As a result, both giving roars to competitors and judging an opponent on the basis of his roars have become evolutionarily stable. From the signaler's perspective, it's always better to roar than to remain silent: if you're a large male, you can win the dispute without risking a fight; if you're a small male, you will avoid costly aggression and the risk of injury and you may, occasionally, deceive your opponent into thinking you are larger or more powerful than you actually are. From the listener's perspective, judging a male's relative competitive ability according to the quality of his roars provides an accurate measure of his fighting ability without incurring the potential costs of physical confrontation. Because selection favors recipients who attend only to those features of displays that are correlated with the signaler's competitive ability, signals that are not honest indicators will disappear and the displays that persist will be generally truthful, or "honest on average" (see Grafen 1990a, b for theoretical details; Searcy and Nowicki 2005 for a full review; Laidre and Johnstone 2013 for a recent useful treatment).

The honest signaling hypothesis assumes that signals function as they do because one individual's display provides another with *information*, defined as a reduction in the recipient's uncertainty about the signaler or what the signaler will do next (Beecher 1989; Seyfarth et al. 2010). Signals are informative whenever there is a predictable relation between the signal and current or future events, thereby reducing the recipient's uncertainty about what is likely to happen next. No special cognition is required. Simple Pavlovian conditioning could suffice: a tone predicts shock, not food; an alarm call predicts an eagle, not a leopard; an individually distinctive scream predicts that animal X, and not animal Y, is involved in a dispute (Seyfarth and Cheney 2003; Seyfarth et al. 2010).

Communication during Cooperative Interactions

Does the honest signaling hypothesis—developed originally to explain the ubiquity of competitive displays—also apply to the many other signals that animals use in more cooperative circumstances? As Searcy and Nowicki (2005) point out, whenever two animals come together there is uncertainty about the outcome, because the best strategy for one depends upon what the other does, and vice versa. Communicative signals have evolved, at least in part, to resolve this uncertainty.

Consider, for example, the grunts given by female baboons when they attempt to interact with another female's newborn infant (Silk et al. 2003). All females are attracted to young infants, but mothers are sometimes reluctant to allow their infants to be handled and avoid other females' approaches, particularly those of higher-ranking females (Cheney, Seyfarth, and Silk 1995a). The interaction has an uncertain outcome because neither female knows what the other's response will be. Field observations (Cheney, Seyfarth, and Silk 1995b) suggest that grunting by the approaching female reduces this uncertainty, because friendly interactions are much more likely to occur when she grunts than when she does not.

Silk, Kaldor, and Boyd (2000) tested this hypothesis in a study of rhesus macaque (*Macaca mulatta*) females, who often give grunts or "girney" vocalizations as they approach mothers with infants. They found that these vocalizations did, indeed, predict an approaching female's subsequent behavior. If she gave a vocalization, she was significantly less likely to be aggressive, less likely to elicit submissive behavior from the mother, and more likely to groom the mother than if she remained silent.

There was, in other words, a contingent, predictable relation between the approaching female's vocalizations and what she did next. Anxious mothers had learned to recognize this relation; they acquired *information* from the approaching females' vocalizations because the vocalizations reduced their uncertainty about what was likely to happen next. The calls had acquired *meaning*. The grunts and girneys of baboons and macaques provide one of many examples in the animal kingdom where "under favorable conditions, unsophisticated learning dynamics can spontaneously generate meaningful signaling" (Skyrms 2010:19).

Silk, Kaldor, and Boyd (2000) then generalized this result using a model that showed that honest, low-cost signaling can evolve even when there is some degree of conflict between the animals involved. They argued further that such signaling is particularly likely to become evolutionarily stable when coordination between partners is valued and animals interact repeatedly over time. This conclusion is important because these are just the conditions that exist in most primate groups, and indeed in many other groups of birds and mammals.

The Social Function of Nonhuman Primate Vocalizations

Given that animal signals are generally honest predictors of the signaler's subsequent behavior, that they reduce uncertainty in listeners, and that they acquire meaning and thereby facilitate social interaction, we are now in a position to evaluate the function of vocal communication among nonhuman primates in light of Clark's (1996) propositions describing the function of human language. Our analysis focuses on a long-term study of wild baboons (*Papio cynocephalus ursinus*) living in the Okavango Delta of Botswana. Like Clark (1996:4), we begin with a tour through the settings of communicative use, the individuals who play a role in these settings, and the way joint actions emerge from individual actions.

Baboons live throughout the savannah woodlands of Africa in groups of 50 to 150 individuals. Although most males emigrate to other groups as young adults, females remain in their natal groups throughout their lives, maintaining close social bonds with their matrilineal kin. Females can be ranked in a stable, linear dominance hierarchy that determines priority of access to resources. Daughters acquire ranks similar to those of their mothers. The stable core of a baboon group is therefore a hierarchy of matrilines, in which all members of one matriline (for example, matriline B) out-

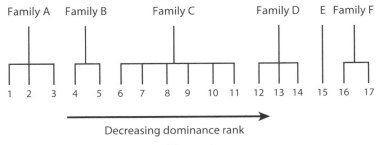

Figure 1

rank or are outranked by all members of another (for example, matrilines C and A, respectively). Ranks are extremely
stable, often remaining unchanged for decades (Silk, Altmann, and Alberts 2006a, b; Cheney and Seyfarth 2007).

When rank reversals occur within a matriline, they affect
only the two individuals involved. However, when rank reversals occur between individuals in different matrilines,
most members of the lower-ranking matriline rise in rank
together above the members of the previously higher-ranking
matriline (Cheney and Seyfarth 2007). Figure 1 illustrates the
hierarchical, matrilineal society of baboons.

Baboon vocalizations are individually distinctive (e.g.,
Owren, Seyfarth, and Cheney 1997), and playback experiments have shown that listeners recognize the voices of others as the calls of specific individuals (reviewed in Cheney
and Seyfarth 2007). The baboon vocal repertoire contains a
number of acoustically graded signals, each of which is given
in predictable contexts (e.g., Fischer et al. 2001). For example, grunts—the baboons' most common vocalization—
may be given to either a higher-ranking or a lower-ranking
individual. By contrast, screams and fear-barks are given
almost exclusively by lower-ranking to higher-ranking animals, whereas threat-grunts are given only by higher-ranking
to lower-ranking individuals. Because calls are individually
distinctive and each call type is predictably linked to a par-

ticular social context, baboon listeners can potentially acquire quite specific information from the calls that they hear.

A Listener Recognizes the Caller's Intention to Communicate *to Her*

Baboon groups are noisy, tumultuous societies, and an individual would not be able to feed, rest, or engage in social interactions if she responded to every call as if it were directed at her. In fact, baboons seem to use a variety of behavioral cues, including gaze direction, learned contingencies, and the memory of recent interactions with specific individuals to make inferences about the target of another individual's vocalization. For example, when a female hears a recent opponent's threat-grunts soon after fighting with her, she avoids the signaler—that is, she acts as if she assumes that the threat-grunt is directed *at her*. However, when she hears the same female's threat-grunts soon after grooming with her, she shows no such response, acting as if she assumes that the calls are directed at someone else (Engh et al. 2006).

The attribution of motives is perhaps most evident in the case of "reconciliatory" vocalizations. Like many other group-living animals, baboons incur both costs and benefits from group life. In an apparent attempt to minimize the disruptive effects of within-group competition, many primates "reconcile" with one another after aggression, by coming together, touching, hugging, or grooming (Cheney, Seyfarth, and Silk 1995a). Among female baboons, reconciliation takes place after roughly 10 percent of all fights, and typically occurs when the dominant animal grunts to the subordinate within a short time after aggression. Playback experiments have shown that, even in the absence of other behavior, such reconciliatory grunts increase the likelihood that the subordinate will approach her former opponent or tolerate the opponent's approach. By contrast, playback of a grunt from another dominant, previously uninvolved individual has no such effect (Cheney and Seyfarth 2007). Baboons in these experiments behaved as if hearing a reconciliatory grunt from their former opponent affected their assessment of the opponent's intentions toward them.

Call Meaning Depends on
Multiple Sources of Information

Baboons' responses to reconciliatory grunts appear to depend on the listener's integration of information from the call type, the caller's identity, and the listener's memory of previous interactions with the caller. Further experiments suggest even greater complexity. For example, Wittig, Crockford, Ekberg, et al. (2007) found that baboons will accept the reconciliatory grunt by a close relative of a recent opponent as a proxy for direct reconciliation by the opponent herself. When a female, say, D1 (where letters are used to denote matrilines and numbers the individuals within each matriline), received a threat from female A1 and then heard a grunt from female A2, she was subsequently more likely to approach, and more likely to tolerate the approaches of A1 and A2 than if she had heard no grunt or a grunt from another high-ranking individual unrelated to the A matriline. Intriguingly, D1's behavior toward other members of the A matriline did not change.

Similarly, in a test of vocal alliances among baboons, a subject who had recently been threatened by a more dominant female heard either the aggressive threat-grunt of a close relative of her opponent or the threat-grunt of a female belonging to a different matriline. Subjects responded more strongly in the first condition, avoiding both the signaler, the original antagonist, and other members of their family for a significantly longer time than in the control condition (Wittig, Crockford, Seyfarth, et al. 2007). In both of these experiments, subjects responded as if the meaning and intended target of a call, and hence their response to it, depended not only on the caller's identity and the type of call given but also on the caller's kinship relationships with any individuals with whom the subject had recently interacted. Subjects acted as if they attributed some kind of shared intention—to reconcile, or to form an alliance—to closely bonded individuals, a shared intention that they did not attribute to individuals who belonged to different matrilines.

COMMUNICATION REVEALS WHAT BABOONS KNOW ABOUT EACH OTHER

Baboons are born into a social world that is filled with statistical regularities: animals are recognized as individuals and interact in highly predictable ways. Not surprisingly, baboons recognize these regularities, particularly those associated with dominance rank and the strong bonds among matrilineal kin. In playback experiments, for example, listeners respond with apparent surprise to sequences of calls that appear to violate the existing dominance hierarchy (Cheney, Seyfarth, and Silk 1995a). In other experiments, they demonstrate knowledge of other animals' matrilineal kin relations by, for instance, looking toward the mother when they hear an offspring's scream (Cheney and Seyfarth 1999), or by accepting as reconciliation a grunt from a close relative of a recent opponent as a proxy for reconciliation with the opponent herself (Wittig, Crockford, Ekberg, et al. 2007).

To test whether baboons integrate their knowledge of other individuals' kinship and rank, Bergman et al. (2003) played sequences of calls mimicking rank reversals to subjects in matched trials. In one set of trials, subjects heard an apparent rank reversal involving two members of the same matriline: for example, B3 threat-grunts and B2 screams. In another set, the same subjects heard an apparent rank reversal involving the members of two different matrilines: for example, C1 threat-grunts and B3 screams (recall that in baboon society a within-family rank reversal affects only the two individuals involved, whereas a between-family reversal is a more momentous social event that affects all of the members of both matrilines). Between-family reversals elicited a consistently stronger response than did within-family rank reversals (Bergman et al. 2003). Subjects acted as if they classified individuals simultaneously according to both kinship and rank. The classification of individuals simultaneously according to two different criteria has also been documented in Japanese macaques (Schino, Ventura, and Troisi 2005).

Continuities in Social Function

From these data it should be clear that, despite their considerable differences in structure and communicative power, human language and nonhuman primate vocalizations share many similar functions.

- *Both are used for social purposes.* Like humans, monkeys and apes use communication to facilitate social interactions, maintain group cohesion, and alert others to food and predators.
- *Both constitute a form of joint action* in which two or more individuals, in socially defined roles, carry out and coordinate their activities (Clark 1996). These include, for nonhuman primates, aggression, competition, cooperation, reconciliation, and the formation of alliances.
- *Both are used primarily in face-to-face interaction.* Although discussions of the evolution of language have paid considerable, often exclusive attention to the predator alarm calls given by monkeys (e.g., Bickerton 1990), these constitute a relatively rare and perhaps highly specialized form of communication in animals with few parallels in human language. Alarm calls are useful in studies of the "meaning" of animal signals (e.g., Zuberbuhler, Cheney, and Seyfarth 1999), but they may be of limited use when thinking about the evolution of language. By contrast, the most common vocalizations used by nonhuman primates are the grunts, girneys, chutters,

threat-grunts, and screams used during close-range, face-to-face interactions, and these calls have many functional parallels with human conversation.

- *Both involve speaker's meaning and addressee's understanding.* In both language and nonhuman primate communication, we can deconstruct any communicative act between two individuals and examine separately what a signaler intends and what information a recipient acquires (Seyfarth and Cheney 2003). In both systems, signalers are motivated to affect the behavior of recipients, while recipients acquire information about the signaler's likely behavior. For recipients in both systems, the meaning of a call derives from a variety of sources (what linguists call pragmatics), including the call type, the caller's identity including her rank and family membership, and the recipient's memory of past events.

This said, there are also striking differences between language and nonhuman primate communication in the meaning of calls to signaler and recipient. Among nonhuman primates, these two sorts of meaning are sometimes the same, for example, in cases where a female baboon's grunt signals her motivation to act in a friendly manner and the listener, recognizing the predictive relation between grunting and friendly behavior, relaxes. In other cases, however, meaning to signaler and recipient are strikingly different: for example, in cases where a female baboon gives a "contact bark" when she is separated from the rest of her group. Although the bark provides listeners with information about her location, playback experiments suggest that the caller's behavior is motivated primarily by her current state of separation rather than her intention to inform others (Cheney, Palombit, and Seyfarth 1996). This highlights a crucial difference between language and nonhuman primate communication: a difference that derives from the signaler's and listener's ability (or inability) to represent the mental state of the other. As Fitch (2010) points out, human language involves a cognitive parity between speakers and listeners, where both know the meaning of a word, and know that the other knows this. Nonhuman primates may exhibit a rudimentary theory of mind—they may know, for example, that another's call is directed at them (see above), or that another individual does not know about a nearby snake or newly discovered food (Crockford et al. 2012; Schel, Machanda, et al. 2013; Schel, Townsend, et al. 2013)—but there is no evidence that participants in these interactions have the kind of cognitive parity routinely found in language.

Of course, similar functions do not mean that language and nonhuman primate vocalizations are identical modes of communication. Obviously, enormous changes in brain size, communication, and cognition have taken place since the divergence of the human and nonhuman primate lineages roughly six million years ago. The differences between language and nonhuman primate communication are well known, widely discussed (Hurford 2007; Fitch 2010), and need not be repeated here.

By contrast, far less attention has been paid to the many parallels between the function of language in our daily lives and the function of vocalizations in the daily lives of nonhuman primates. These functional similarities are important because they remind us that, for all their structural differences, human and nonhuman primate communication are overwhelmingly social phenomena, designed to facilitate interactions between individuals, many of whom have known each other and interacted socially for years. More important, these common social functions help explain why human and nonhuman primates share so many homologous neural mechanisms that control the perception, integration, and comprehension of visual and vocal signals. We turn now to the homologous cognitive mechanisms that underlie the two systems—that is, to Clark's (1996) final proposition, that *the study of language use is both a cognitive and a social science*.

CONTINUITIES IN COGNITION

It is now well accepted that when natural selection acts to favor a particular behavior, it simultaneously acts to favor whatever brain mechanisms and cognitive abilities are required to make that behavior possible. For example, in the black-capped chickadee (*Poecile atricapillus*), a bird that stores seeds in the fall and retrieves them in winter, increasingly severe winters from Kansas to Alaska have led to an increase not only in caching and retrieval behavior but also in hippocampal neuron number, neurogenesis, and spatial memory (Pravosudov and Clayton 2002; Roth and Pravosudov 2009; Roth et al. 2011). If this principle holds in the case of nonhuman primate communication and human language, we might expect that similar social pressures have led not only to homologous brain structures but also to at least some similar cognitive skills. The prediction deserves attention because the cognitive skills that underlie language—a discrete, combinatorial, rule-governed, and open-ended computational system—seem so utterly different from the cognitive skills that underlie nonhuman primate communication.

To examine the cognitive mechanisms underlying the perception of nonhuman primate vocalizations, recall the experiment in which Bergman et al. (2003) played to baboon subjects a sequence of threat-grunts and screams from two adult females. As already noted, this experiment was designed to test the hypothesis that individuals classify others simultaneously according to both rank and kinship, and would therefore respond more strongly to call sequences that mimicked a between-family rank reversal than to those that mimicked a within-family rank reversal.

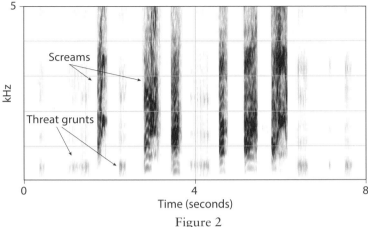

Figure 2

Subjects' responses indicated not only that they recognized when a rank reversal had occurred but also that they recognized that some rank reversals were potentially more significant than others. Upon hearing a scream-threat-grunt sequence, baboons appeared to form a mental representation of a specific social interaction involving specific individuals. This representation developed instantly (within seconds) and was built up from several discrete pieces of information: the type of call, the caller's identity, and the caller's rank and kinship affiliation. In the experiments just described, the threat-grunt is recognized as an aggressive call produced by a particular individual (say, Hannah), who is further identified as belonging to a middle-ranking matriline. The scream is recognized as a distress signal (and not some other call) coming from, say, Sylvia, a member of the highest-ranking matriline. We know that each acoustic element in the call sequence carries its own independent meaning because if we change the identity of either caller or the type of call given, listeners respond differently. Figure 2 represents a typical playback sequence of threat-grunts and screams.

These discrete elements, moreover, are combined according to the "rules" of call delivery to create a message whose meaning is more than just the sum of the meanings of its constituent elements. Listeners interpret the sequence of calls as a narrative in which Hannah's threat-grunts *are directed at* Sylvia and *cause* Sylvia's screams. Without this element of causality—if the calls were perceived as juxtaposed simply by chance—there would be no violation of expectation and no strong response to the apparent rank reversal.

Of course, in many respects, the call sequence could hardly be more different from language: it is produced by two individuals, one of whom is threatening and the other is screaming. But in the mind of the receiver, the string of elements is organized in what one might call a narrative account, with an actor, an action, and an acted upon. And it is a narrative account that would have a different meaning—and elicit a different response—if we changed any one of its elements. It is also a narrative account that can be judged by the listener as either true or false—that is, either consistent or not consistent with what the listener knows about her social group.

The baboons' assessment of call meaning thus constitutes a *discrete, combinatorial, and rule-governed* system of communication (Cheney and Seyfarth 1998; Worden 1998) in which a finite number of signals can yield a nearly infinite number of meanings. If a listener recognizes the difference between [A threatens B and B screams] and [B threatens A and A screams], and can make this distinction for every dyad in a group of 70–80 individuals, a simple system of signals can generate a huge number of meanings. Indeed, the communicative system is effectively open-ended, because baboons learn to recognize the calls of new infants born into their group, or of new male immigrants, and assign meaning to these calls depending on the individuals' ranks and kinship affiliations. Because the rules of assignment can be applied to *any* new individual, the potential number of different meanings is completely open-ended.

We make these points not to argue that vocal communication in baboons constitutes a language, or even to claim that baboon communication has many of language's formal, structural properties. Instead, we suggest that several of the cognitive mechanisms that have long been thought to mark a clear separation between language and nonhuman primate communication can, in fact, be found—in admittedly simpler form—in the communication and social cognition of nonhuman primates. As a result, the earliest steps toward the evolution of language may not be as difficult to imagine as Pinker (1994) first proposed. Instead, just as we find in the anatomy of quadrupedal chimpanzees morphological features that very likely served, in the common ancestor of chimpanzees and humans, as precursors to human bipedalism, so we can find in the communication of nonlinguistic baboons cognitive operations that very likely served, in our mutual common ancestors, as precursors to the cognitive operations that underlie language.

Why should such a system have evolved? Long-term field studies demonstrate that the best predictor of a baboon's or chimpanzee's reproductive success is an individual's ability to form close, long-term bonds and to recognize the relations that exist among others. Among female baboons, for example, Silk, Alberts, and Altmann (2003) and Silk et al. (2009, 2010) found that the best predictor of an individual's offspring survival and longevity was the strength of her social bonds with other females—usually, but not always, close matrilineal kin. Among male chimpanzees, the best predictor of a male's lifetime reproductive success was his ability to form coalitions with other males, and the greatest beneficiaries of coalitionary aggression were those individuals "who tended to have coalition partners who themselves did not form coalitions with each other" (Gilbey et al. 2012:373). Selection has thus favored, in both species, individuals who are skilled not only in the use of communication to form and maintain bonds but also in the ability to derive, remember, and make use of information about other animals' relations. Discrete, combinatorial communication and cognition has thus been favored by natural selection.

SOCIAL KNOWLEDGE AS A COGNITIVE PRECURSOR OF LANGUAGE

In many respects our proposal is not new. In 1990, for example, Pinker and Bloom suggested that during the course of human evolution, "grammar exploited mechanisms originally used for . . . conceptualization" (1990:713), while Newmeyer (1991:10) argued that "the conditions for the subsequent development of language . . . were set by the evolution of . . . conceptual structure. A first step toward the evolution of this system . . . was undoubtedly the linking up of individual bits of conceptual structure to individual vocalizations" (for similar views, see Jackendoff 1987, 2002; Kirby 1998; Newmeyer 2003; Hurford 1998, 2003). Our proposal is new, however, in its emphasis on social cognition and in our ability to link social cognition with reproductive success.

Three sorts of cognition, all well documented in animals, have been offered as possible cognitive precursors of language (Hauser, Chomsky, and Fitch 2002): systems of orientation and navigation (e.g., Menzel 2011; Jacobs and Schenk 2003); number (Feigenson, Dehaene, and Spelke 2004; Cantlon and Brannon 2007; Jordan, MacLean, and Brannon 2008); and social cognition (Worden 1998). All involve discrete elements and rule-governed computations. All help explain how the discrete, combinatorial structure found in language might have evolved from some prelinguistic cognitive precursor. In at least three respects, however, social cognition seems the most likely candidate as a precursor of language.

First, only in social cognition do the discrete elements include both living creatures, to which listeners can reasonably attribute motives and goals, and context-specific vocalizations, each associated with a caller's motivation to interact with another in specific ways. As a result, a crucial part of primate social cognition involves the attribution of intent to others, therefore constituting a rudimentary form of propositional thinking in which there is a subject and a predicate ("Sylvia is mad at Hannah"). This distinguishes social cognition from orientation, navigation, and number, where there are no agents, actions, and patients. In his discussion of the possible evolutionary origins of language, Hurford (1990, 2007) asks whether propositional structures are unique to language, or whether they "somehow existed before language" in another domain. Social cognition provides an answer. The propositions that are expressed in language did not originate with language—they arose, instead, because to succeed in a social group of primates one must engage in an elementary form of propositional thinking.

Second, only in social cognition are the discrete elements explicitly linked to vocalizations, so that the system of communication and the system of cognition on which it is based are tightly coupled. This merging of cognition with communication does not occur in orientation, navigation, or number comprehension.

Third, only in social cognition are the discrete elements linked—as in language—to the organization of knowledge into concepts. When baboons hear a sequence of vocalizations that violates the dominance hierarchy, they respond within seconds. When a male macaque involved in a fight tries to recruit an ally, he seems instantly to know which individuals would be the most effective partners (Silk 1999). The speed of these reactions suggests that animals are not searching through a massive, unstructured database of associations but have instead organized their knowledge of other individuals into concepts: what we call dominance hierarchies and matrilineal (family) groups. These social categories qualify as concepts for at least two reasons. First, they cannot be reduced to any one, or even a few, sensory attributes. Family members do not always look alike, sound alike, behave alike, or share any other physical or personality features that make them easy to tell apart (Seyfarth and Cheney 2013, 2014). Second, social categories persist despite changes in their composition. Among females and juveniles, the recognition of families is unaffected by births and deaths; among adult males, the recognition of a linear, transitive hierarchy persists despite frequent changes in the individuals who occupy each position (Kitchen, Cheney, and Seyfarth 2003). In the mind of a baboon, therefore, social categories exist independent of their members. And because the meaning of a vocalization cannot be divorced from the caller's identity, and the caller's identity cannot be separated from her placement in a conceptual structure based on kinship and rank, communication and conceptual structure are inextricably bound together, just as we might expect in a system of communication that served as a precursor to human language and thought.

Human language and nonhuman primate vocal communication share many social functions. Both constitute a form of joint action and are used for social purposes in face-to-face interactions. Both help to reduce uncertainty, regulate social interactions, and establish and maintain social relationships. Given these common functions, it is not surprising to find that homologous neural mechanisms underlie both systems: mechanisms for the recognition of individual faces and voices, for cross-modal integration, and for the recognition of objects and the assessment of call meaning.

Finally, despite their many well-established differences, language and nonhuman primate communication share a suite of common cognitive operations. Both are discrete, combinatorial systems in which a finite number of signals can generate an infinite number of meanings. In both systems, it is possible to distinguish the speaker's meaning from the addressee's understanding. And in both systems the meaning of a signal is deeply embedded in the social context: meaning depends not only on the signal itself but also on the caller's and listener's identities, and their history of interaction with each other and related individuals.

The simplest way to explain these shared traits is to propose that during their long, common evolutionary history, the ancestors of modern monkeys, apes, and humans faced similar social problems and evolved similar systems of communication to deal with them. Although human language subsequently evolved to become markedly different from nonhuman primate communication, their shared evolutionary history is apparent in homologous neurophysiology, social function, and cognition. The earliest stages of language evolution—long thought to be a mystery—are easier to imagine if we focus on the social function of communication in a complex society where individuals use communication and cognition to manage their social relations and represent the relations of others. Long before it became language, primate communication and cognition were social devices, designed by natural selection to achieve certain goals. Long after it evolved, language still is.

used not only to mean "bearing hope"—"He operated hopefully"; an objective, descriptive usage—but also to indicate that the speaker has a hopeful attitude toward a proposition. "Hopefully, he will be here" is a subjective, *pragmatic* usage, now more common than the supposedly "proper" one. *Rather* began with the objective meaning "earlier" and only eventually evolved to mean "preferably" as a metaphorical extension of "earlier" (that which is done earlier is likely that which one finds more compelling). The "preferably" meaning is attitudinal, and therefore pragmatic.

Yet, research on grammaticalization remains engaged most on the basis of the documentation of the emergence of things more traditionally considered to be "grammar," such as the emergence of the *gonna* future in English from what began as a literal usage of *going to*. The pragmatic work gets less attention.

I suspect that even linguists are subtly discouraged from dwelling at length on such work for broad societal reasons. The fact that pragmatic usages in language are often distrusted by prescriptive attitudes as "vague" or incorrect, combined with the fact that pragmatics is not taught in schools and rarely in linguistics classes, cannot help but encourage a sense of this work on historical pragmatics as somehow subsidiary. Real grammar, one may suppose in the depths of one's linguistic soul, is preterites and coronals and agents and patients—not sociality, attitudes, and cooperation.

LANGUAGES AND PRAGMATICS

Rather, the view from the linguist's desk makes it seem plausible to assume that the birth of language was a matter of developing the ability to construct propositions—statements about things. That human language is distinct in its rich combinatorial possibilities seems, naturally, to deserve the most focus. Take the utterance "Betty is here": conveying that something of note about Betty is that she is here requires this combinatorial substrate, without which it could only convey *Betty* and *here* in unassociated sequence. Hence the focus in Chomskyan research on the fundamental operations of Move and Merge, and the associated interest in recursion as supposedly unique to human language (Hauser, Chomsky, and Fitch 2002). Related is Bickerton's work on protolanguage (1990, 2009), in which a level of expression equivalent to that in primitive pidgins is transformed, via the emergence of operations of the Move and Merge variety, into a full syntax.

After several tens of millennia of existence, much of what languages constitute would indeed seem to be syntax plus phonology and morphology: "grammar" in the classic sense. One thinks of a Russian sentence such as:

Ja uvid-el i po-kup-il   krasn-uju knig-u.

I see-PAST and PERF-buy-PAST red-ACC book-ACC

"I saw and bought the red book."

It is hardly unreasonable to suppose that unraveling the emergence of a capacity that can produce and process sentences such as these, with its rich morphology riddled with complexities and irregularities to be mastered, has all to do with grammar and little of interest to do with getting along. Certainly, one thinks, this machinery would not have emerged in a species that was *not* a social one: a sentence like the one above is used between people, not internally by one person. Yet, the mechanical distractions of such a sentence may make work proposing social origins of language such as Dunbar's (1998) on gossip, Mithen's (2005) on song, and Falk's (2009) on motherese seem, at best, previous to when "the real stuff" happened, as if one traced a hurricane not to an emergent weather pattern but to the very emergence of H_2O and wind currents amid the formation of the planet earth.

MODERN DEVELOPMENTS AND PRAGMATICS

Seyfarth and Cheney's argument corresponds to growing evidence that a serious theory of linguistic evolution must put the "social" first. I do not mean this in the sense often leveled by sociolinguists—that a theory of language that does not take social context into mind is logically incomplete. This claim, when leveled at work on phonology, syntax, and semantics, is often a rather airy one. However, promising newer theories of grammar (Jackendoff 2002; Culicover and Jackendoff 2005) insist that evolutionary plausibility requires that Chomskyan "syntactocentrism" be upended in favor of a model in which semantics came first and still precedes syntax in the mental mechanism that generates language. It seems increasingly advisable that pragmatics precedes even semantics in such models, both in the sense of synchronic language as well as how language emerged, despite the temptations of seeing pragmatics as something added on after the "real" work has happened.

Scott-Phillips (2015) argues, for example, that what distinguishes human communication is not simply referentiality and combinatoriality, both of which are seen in animal communication systems and which, if evolutionarily advantageous in expanded form, would surely have exploded in creatures other than us long before now. Rather, he posits that human language is distinct in being founded on a capacity to generate and process the very intention to communicate (his term is *ostensive* communication) beyond the rudimentary level of a few symbolic totems such as warnings or directions to food.

Scott-Phillips bases his ideas on principles familiar to philosophers of language such as establishing relevance as well as distinguishing that which is urgent from that which is not, essentially making an argument that pragmatics—what he titles pragmatic competence—established a substrate upon which what we think of as grammar developed. Moreover,

Scott-Phillips argues that this "grammar" can only have developed on the basis of the ostensive substrate, without which there would be no communicative imperative to encourage such a baroque machinery to evolve and be evolutionarily selected for.

Linguists researching the almost counterintuitively extreme complexity of most human languages have been converging on evidence that confirms the premise of work such as Scott-Phillips and Seyfarth and Cheney. A sentence such as the Russian one above evidences a great deal of elaboration. It is becoming increasingly clear, however, that this quality emerges in language not because of any correspondence to cultural imperatives or functional necessity (contra Lupyan and Dale 2010) but because languages passed down generations learned only by children rather than adults amass complexity simply by chance. Languages function equally well without such machinery, as becomes clear from creole languages, which emerge when a rudimentary pidgin (or pidgin-like) variety is expanded into a full language. Such a language, having recently been a pidgin with minimal grammatical machinery and not having existed long enough to "rust up" with unnecessary but processable complexities, is inevitably more streamlined in the grammatical sense than a language like Russian. Yet there is no evidence that such languages, nor other languages in the world less radically streamlined than creoles but close to them in the scalar sense of the matter, prevent nuanced communication.

More to the point, in the rare cases when language emerges anew from what was not language, such as creoles and sign languages, the new language is never riddled with complexity like Russian. New language is, in its way, sensible language. Time mucks things up. The lesson is that "grammar" as traditionally known is, to a considerable degree, an accidental development despite its complexity. As such, a theory of language evolution that assumes the origin of a structure like Russian is what we must explain is misguided: the "language" that emerged at first was surely much more like a creole than like Russian.

Under such a perspective, a space opens up within which morphophonemics and future tense paradigms occupy less central a space in what we consider "language" to be. This is already plain in languages that do not happen to have the rich affixal machinery familiar from Indo-European languages on which modern linguistics was founded. A linguistics that proceeded from, for example, East and Southeast Asian languages would take pragmatics as much more fundamental.

Here, for example, is a sentence of Cantonese, where the final four words are pragmatic particles that are key in conveying the meaning of the sentence. "She got first place" as a faceless utterance is not, in truth, the meaning of such a sentence; central to its meaning is why the utterance is being made, and in Cantonese, the pragmatic machinery getting that across is fundamental, with each particle conveying a unit of attitudinal meaning:

Kéuih ló-jó daih   yāt mìhng **tìm gela**   **wo**   .

she   take-PERF number one place too PRT PRT PRT

"And she got first place too, you know." (Matthews and Yip 1994:345) (*tìm* is evaluative, *ge* is assertive, *la* denotes currency, *wo* newsworthiness)

There are dozens of such particles in Cantonese, which is typical of languages of that part of the world. Using them is basic to speaking the language as a person rather than as a learner or automaton. A fair observation is that this proliferation of pragmatic particles is typical of languages low on affixal morphology; it would seem that languages like Indo-European ones are much less likely to have this many overt and regularized pragmatic items. Possibly, factors of economy dictate that a language leave more of these nuances to context or intonation as affixal morphology clutters up the typical utterance string. However, languages such as Cantonese make clear that there are no grounds for supposing that "grammar" in the Indo-European sense is somehow default, and possibly even support pragmatics as more fundamental than what most linguists consider syntax to consist of.

A similar conclusion may be gleaned from my own work on the emergence and development of the creole language Saramaccan since the late seventeenth century. Pragmatics has been as central to the emergence of the language's central expressive machinery as morphology and syntax. In fact, Saramaccan seems to be in no special hurry to develop affixal morphology; however, it has developed pragmatic machinery with what seems almost urgency.

Nɔ́ɔ a bái wǎ hási seéi.

ʔ he buy a horse ʔ

"So, he even bought a horse."

In this sentence, the first and final words are of a kind often marked as just "emphatic" or roughly translated as meaning vague things like "then." However, they are pragmatic markers, with precise functions: *nɔ́ ɔ* indicates that a sentence imparts new information rather than old, assumed information, while *seéi* conveys that something is counterexpectational in various ways; "even" is the best translation here. These words in the language have attracted little study (an exception is McWhorter 2009) in favor of examinations of Saramaccan's "real" grammar. However, the sense that its ways of conveying past tense and marking definiteness are somehow "realer" than this battery of pragmatic markers, of which I have exemplified but two, is artificial. Saramaccan demonstrates that the emergence of a true language from a pidgin entails, quite preliminarily, the establishment of these socially oriented mechanisms. In turn, this helps make a socially based origin scenario for language seem much more plausible.

In short, a linguist can see examinations of language origins based on social interactions as falling outside of what makes language interesting. However, increasing evidence can be interpreted as showing that mediation between the competing desires and attitudes of persons is as basic to constructing real sentences as marking verbs for tense or nouns for plural. As such, work such as Seyfarth and Cheney's has more to do with what we do than may always be truly appreciated.

2

WHERE IS CONTINUITY LIKELY TO BE FOUND?

Ljiljana Progovac

In part 1 of this volume, Seyfarth and Cheney report on some important results regarding the cognition of baboons by focusing on specific and falsifiable hypotheses, to the effect that, very generally speaking, baboons are able to compute the appropriateness of submission and threat calls. The authors extrapolate from this finding and draw conclusions about the evolution of human language. They are right to point out that the differences between human language and nonhuman primate communication are widely discussed and well known, while much less attention has been devoted to the potential points of contact, or continuity, between the two kinds of systems. Here I offer my own observations regarding their findings concerning the cognition of baboons and address how these and comparable findings about animal communication can be relevant to understanding the evolution of human language. Linguists need to meet biologists, anthropologists, and other researchers of language evolution (at least) halfway with specific theories of language evolution. To that end, I present a specific, linguistically based proposal about the evolution of syntax/grammar, in an attempt to reveal where continuity between human language and the abilities of the other primates can be sought.

My conclusion is that continuity should be sought in the most basic constructions of human language rather than in the most elaborated. In order to identify what these are, one needs an explicit theory of language evolution. I also maintain that continuity should be sought not only in the aspects of human language that can be found (or not) in animal communication systems but also in the opposite direction, in the aspects of animal communication systems that can be found in human language. With an elaborated theory of language evolution, one has a much more grounded platform from which to formulate specific and falsifiable hypotheses about continuity in both directions. While there is a great deal of good research, including Seyfarth and Cheney's, on animal communication systems, there is relatively little specific and focused research by linguists on the topic of language evolution.

BABOONS' COGNITIVE ABILITIES AND LANGUAGE EVOLUTION

Seyfarth and Cheney report on experiments with female baboons in Botswana, who seem to be able to mentally compute the appropriateness of submission and threat calls. The individuals of higher rank typically issue threat calls toward those of lower rank, but not vice versa. Likewise, the individuals of lower rank typically issue submission calls toward those of higher rank, but not vice versa. The experiments consisted of recording and playing back both the expected and the reversed calls, and established that the baboons react by paying more attention to the call reversals than to the expected calls. In addition, Seyfarth and Cheney report that baboons' responses to reconciliatory grunts show great complexity, as they take into account not only the caller's identity and the call type but also the listener's memory of previous interactions with the caller. Their book *Baboon Metaphysics* (2007) provides thorough and lucid background, including how ranking is established among females (the group studied in this experiment), how the baboons know who has a higher or lower status than them, and how the researchers determine this. The book also elucidates how and why rank reversals sometimes occur.

These are significant findings, as they show that quite complex reasoning can take place without (human) language. The question is, then, what human language brings about that improves on this ability, or some other abilities that are relevant for selection, as well as what the concrete steps that brought about human language were.

As I see it, the reported results seem to disconfirm one particular stance regarding language evolution, in which a similar issue is raised: that of a relationship between language and thought. Chomsky (2010) and some other saltationists (e.g., Berwick and Chomsky 2011, 2016; Bolhuis et al. 2014) basically equate language with thought, proposing that language evolved separately from its externalization. According to this view, language emerged (in full) to facilitate thought (inner speech), rather than communication; once this thought system was externalized (i.e., pronounced), then it could have proved useful for communication as well. More precisely, according to Berwick and Chomsky (2011:40–41), "in the very recent past, maybe about 75,000 years ago, . . . an individual . . . underwent a minor mutation that provided the operation Merge," which brought about recursive structured thought. It was at some later stage that the language of thought was connected to the external speech, "quite possibly a task that involves no evolution at all."[1] It strikes me that their idea of recursive structured thought is directly relevant for Seyfarth and Cheney's proposal of "discrete, combinatorial, rule-governed, and open-ended" reasoning capabilities of baboons.

Now, if Seyfarth and Cheney's interpretation of the results from baboons is correct, then it suggests that one can have "structured thought," if you will, without having language.[2] Language cannot just be equated with thought. Because with humans it is more difficult to separate the two, studies like

Seyfarth and Cheney's on animal cognition are especially relevant as they show quite clearly that thinking and even complex reasoning can take place without (human) language.
While this speaks against the claims made in Berwick and
Chomsky (2011, 2016), one has yet to shed light on how
language actually evolved. What was it about being able to
utter and understand a host of specific words and sentences
that proved advantageous for the human species? And how
could such a progression from no language to a highly complex language take place?

In "The Social Origins of Language," Seyfarth and Cheney
say, "selection has favored in baboons—and, by extension,
other primates—a system of communication that is discrete,
combinatorial, rule-governed, and open-ended. We argue
that this system was common to our prelinguistic primate
ancestors and that, when language later evolved from this
common foundation, many of its distinctive features were
already in place." However, in my view, Seyfarth and Cheney's
results show that baboons have a cognitive system that is
quite sophisticated, and capable of figuring out dominance
ranks based on multiple criteria. But that kind of ability
should be separated from the nature of their communication
system, which seems to consist of a limited number of (noncombinatorial) calls. These calls themselves still seem to just
mean: "I hereby threaten you" or "I hereby submit to you."
That these kinds of simple noncombinatorial calls can interact in complex ways with the cognitive processes and
thoughts of baboons is a separate issue. The vocalizations of
baboons, even though accompanied by impressive cognitive
abilities, do not themselves show combinations of these calls,
or some other audible/visible structures that can be seg-
mented into pieces and then recombined to express different
meanings, which is what characterizes human language and
warrants explanation. Human language has words, thou-
sands of words, which are organized into a system, and into
layers of concreteness versus abstractness, corresponding to

the levels/degrees of grammaticalization. These layers of abstractness are also relevant for the layers of syntactic hierarchical structure.

Talking about "language" as a whole, as well as social factors in general, without focusing on specific aspects of each, makes it hard to formulate falsifiable hypotheses regarding language evolution. For that reason, in my own work, I explore smaller questions and hypotheses about very specific properties of language, such as how human language reached a combinatorial stage, in which two (or more) words are combined into meaningful units; or how human language evolved means for expressing transitivity. Furthermore, I address how and why such specific features of language would have incurred communicative benefits during human evolution. This reconstruction of protosyntactic stages is, moreover, based on a syntactic theory.

SMALL AND SPECIFIC HYPOTHESES

I believe that any continuity with the communication systems of other primates is more likely to be found in the most rudimentary of syntactic structures, rather than in the most elaborated features of human syntax, the latter illustrated below with recursive embedded clauses (1) and recursive possessive determiner phrases (2).

(1) [Mario believes [that Marilyn understands [that Marjorie maintains [that . . .]]]]

(2) [[[[[Mario's] brother's] wife's] uncle's] bicycle] disappeared.

Reasons why recursive structures such as (1) and (2) should be considered highly complex, and late to emerge, are given in Progovac (2015). But how do we know what most rudimentary structures are like? This question is certainly not trivial, and one should avoid impressionistic answers.

My view is that the simpler stages of language can be arrived at only through a precise reconstruction method based on a linguistic theory. I share this view with Heine and Kuteva (2007), whose reconstruction of protolexicon is based on the linguistic framework of grammaticalization. Since grammaticalization typically proceeds in the direction of developing a functional (grammatical) category out of a lexical (content) category (or a more abstract category out of a more concrete category), but not the other way around, Heine and Kuteva reconstruct a stage in the evolution of human language that had content categories (initially only verbs and nouns) but not grammatical categories (e.g., auxiliary verbs, tense markers, articles, subordinators). To illustrate with one example, it is common across languages to grammaticalize verbs meaning "come" into past tense markers, and verbs meaning "go" into future tense markers. The latter is illustrated with English "I *am going* to do this soon," where the verb *go* is used as a future marker, losing its original meaning of physical motion (see Heine and Kuteva 2007 for many crosslinguistic examples and generalizations).

In my own work, I rely on a different linguistic theory to arrive at protostages of grammar. It is encouraging that the two reconstructions (Heine and Kuteva's, and mine) lead to convergent/compatible results, even if they consider different aspects of language and look at them from different angles. What they share is that they both formulate specific and focused research questions, and address them by following a linguistic theory and linguistic data.

The reconstruction that I offer (in Progovac 2015, 2016a, and previous work, e.g., 2009a, b; 2014) leans on the theory of layered syntactic structure associated with Minimalism (e.g., Chomsky 1995) and its predecessors. The simplified hierarchy of functional projections/layers characterizing modern clauses/sentences in this approach is given in (3).

(3) CP > TP > vP > VP/SC

Very roughly speaking, the inner VP/SC (verb phrase/small clause) layer accommodates the verb/predicate and one argument, while vP (light/little verb phrase) accommodates transitivity with one additional argument, such as agent. TP (tense phrase) accommodates the expression of tense/finiteness, while CP (complementizer phrase) accommodates subordination/embedding, for example.[3] This hierarchy is a theoretical construct that offers a natural and precise method of reconstructing previous syntactic stages in language evolution, as given in (4):

(4) Structure X is considered to be primary relative to Structure Y if X can be composed independently of Y, but Y can only be built upon the foundation of X.

While small clauses (SCs)/VPs can be composed without the TP layer, TPs can only be constructed upon the foundation of a SC/VP. Similarly, while SCs/VPs can be composed without a vP layer, the vP can only build its shell upon the foundation of a SC/VP. One can thus reconstruct a TP-less and vP-less (tenseless, intransitive) stage in the evolution of syntax, reduced to only a single layer (SC/VP).

More specifically, my proposal is that the earliest protosyntactic compositions were two-word small clauses, basically flat (not hierarchical, not headed) combinations of a verb-like and a noun-like element, in which the noun, the only argument of the verb, was specified as neither subject nor object. But what might such a grammar look like, and what might be the utility of such a rudimentary grammar? I hypothesized that the grammar behind certain compounds found across languages is an approximation ("living fossil," in the sense of Jackendoff 1999, 2002) of the earliest syntax (see, e.g., Progovac and Locke 2009; Progovac 2009a).[4] The following are some examples of such verb-noun compounds from English (5, 6) and Serbian (7, 8). It is important to emphasize here that I am not claiming that these specific compounds were used in the protosyntactic stage but rather that this kind of two-slot mold, into which comparable creations could be poured, was used in that stage.

(5) pick-pocket, scare-crow, turn-coat, dare-devil, hunch-back, wag-tail, tattle-tale, kill-joy, cut-purse, spoil-sport, saw-bones, burn-house, drynk-pany (drink-penny, miser), pinch-penny (miser)

(6) rattle-snake, cry-baby, stink-bug, worry-wart, copy-cat, tumble-weed, scape-goat, turn-table

(7) cepi-dlaka   [split-hair = hairsplitter]
    deri-koža   [rip-skin = person who rips you off]
    ispi-čutura   [empty-flask = drunkard]
    kosi-noga   [skew-leg = person who limps]
    muti-voda   [muddy-water = one who muddies waters]
    vrti-guz   [spin-butt = fidget]

(8) duri-baba   [sulk-old.woman = who sulks]
    kaži-prst   [show-finger = index finger]
    smrdi-buba   [stink-bug = bug species that smells; one who smells]
    tresi-baba   [shake-old.woman = who shakes/scares easily]
    visi-baba   [hang-old.woman = flower: snowdrop]
    plači-baba   [cry-old.woman = crybaby]
    tuži-baba   [complain-old.woman = tattletale]

Interestingly, while the nouns in the compounds in (5, 7) are object-like (e.g., a killjoy is somebody who kills joy), the nouns in (6, 8) are subject-like (e.g., a crybaby is a baby who cries). This is thus a good approximation of the type of grammar that is incapable of distinguishing subjects/agents from objects/patients grammatically, but instead leaves the interpretation to pragmatics, or to convention. If grammar indeed started in this modest way, then notions such as subject and object would have been irrelevant in this initial stage. The incremental, gradualist approach to the evolution of syntax thus reveals clear and specific communicative advantages of the subsequent stages of syntax, which can, as just one example, provide the means for expressing transitivity and for distinguishing subjects from objects—that is, the participants in an event.

The indeterminacy/underspecification of this kind of grammar is especially obvious in pairs such as *turn-table* and *turn-coat*, which use the same verb *turn*. In the former, *table* is subject-like (table that turns), while in the latter, *coat* is object-like (lit., somebody who turns his coat; traitor), suggesting that the noun position in these compounds is not grammatically specified or predestined to be either a subject or an object. This is also brought to light by the possibility, in principle, to use *turn-coat* to mean "a coat that turns," perhaps a reversible coat, which would be analogous to the interpretation of *turn-table*. For the sake of comparison, more elaborate syntactic structures, such as the compound *head-turn-er*, do not show such indeterminacy, as this compound can only mean somebody who turns heads, not "a head that turns," not even in principle.

Besides the grammatical simplicity of verb-noun compounds, the reader will notice that the majority of these compounds are pejorative, especially the ones that refer to humans, in both English and Serbian. This is true across other languages as well. For a much bigger sample of these compounds, including obscene examples, the reader is referred to Progovac (2015, 2016a) and references therein. Selecting for the ability to quickly produce (and interpret) such (often humorous and vivid) compounds on the spot would have gone a long way toward solidifying not only the capacity to use such simple grammars, the foundation for more complex grammars, but also the capacity for building (abstract) vocabulary. As can be seen in the examples above, these compounds combine basic, concrete words, often denoting body parts and functions, in order to create vivid and memorable abstract concepts (e.g., split-hair, rip-skin, spin-butt, scatter-brain).

According to Progovac and Locke (2009), formation and use of creations comparable to verb-noun compounds may have been an adaptive way to compete for status and sex in ancient times. Their successful use would have enhanced relative status first by derogating existing rivals and placing prospective rivals on notice; and second by demonstrating verbal skills and quick-wittedness. Just imagine the possibilities that would open up for baboons, in their search for status, if they were to acquire this kind of ability. Darwin (1874) identified two distinct kinds of sexual selection: aggressive rivalry and mate choice (see also Miller 2000), both of which seem relevant for the proposed use of this type of compound. Darwin (1872) also pointed out that strong emotions expressed in animals are those of lust and hostility, and that these may have been the first verbal threats and intimidations uttered by humans (Code 2005:322). If so, then we can see some continuity there as well. Discussion of the utility of simple two-word syntax for insult purposes offers but one example of how this would have worked in the context of sexual selection.[5] Whether proven right or wrong, this is a concrete proposal, and it is concrete regarding not only syntactic structure and syntactic data but also the social (communicative) utility of such a structure. As such, it can be subject to testing and falsification.

What is promising about this proposal is that it reveals continuity at two different levels. First, there is continuity of grammar, as these two-slot grammars seem to be within reach to other primates. Second, the pejorative, often obscene nature of these compounds may also reveal continuity with animal communication systems in this respect. Patterson and Gordon (1993) report that the gorilla Koko is capable of producing not only novel compounds but also insult, playfulness, and humor. In fact, it may be instructive in a future experiment with primates to try to teach them some dirty words/signs, and their combinations, to see how motivated they would be to use them. As observed in Darwin (1874), the males of almost all the mammal species use their voices much more during the breeding season, and some are absolutely mute except at this season. If human language was used for display and competitive purposes from the very start, then there is some continuity there, too. Darwin's view in fact was that language evolved gradually through sexual selection, as an instinct to acquire a particular method of verbal display similar to music (see, e.g., Fitch 2010 for reviving these arguments for musical protolanguage).

It is of note that the protosyntactic compounds discussed above often feature swearwords. Code (2005, and references there) provides neurological evidence that swearwords are separately stored from other words, using both the part of the brain where digital language is processed and the part that processes laughing and crying. In that sense, swearwords straddle the boundary between (animal) calls, which share many properties with laughing and crying, on the one hand, and digital language, on the other (see, e.g., Burling 2005). Going back to the point made above, human language does have vocalizations that in some respects resemble animal calls; swearwords are one example. According to Code, such uses of language might represent fossilized clues to the evolutionary origins of human communication, given that their processing involves more ancient patterns, including more involvement of the basal ganglia, thalamus, limbic structures, and right hemisphere.

While we often ask whether we can find characteristics of human language in animal communication systems, we hardly ever pose the opposite question. Given what we know about primate communication systems in their own right, we should be asking if there are constructions in human languages that share characteristics with such systems in some relevant respect. In order to emphasize discontinuity, it is often pointed out that animal communication systems are unlike human language, as they are limited to the here-and-now, and possibly to manipulative acts. The question then arises as to whether human languages have structures that can also be characterized that way. There is no doubt that human languages possess an array of constructions that are used only in the here-and-now and that are manipulative. Besides swearwords, imperatives and exclamatives (such as Duck! Run! Watch out! Oh, no!) are of this nature. This, then, is yet another fertile field for exploring the question of continuity in a specific and tangible way.

CONTINUITY OF GRAMMAR

In seeking continuity with animal communication systems, it is important to look for abilities that can provide a foundation, or a precursor, for certain specific aspects of human language. Given that human language and primate communication systems are not identical, we may not be able to learn much about language evolution by comparing them directly, or by comparing them in toto. My argument is that we instead need to formulate specific and theoretically driven hypotheses about the stages of language evolution and then seek points of continuity involving the simplest stages, while at the same time envisioning how these most basic structures would have provided a platform for introducing more and more innovative structures.

There have been numerous reports that primates are capable of combining two signs into a meaningful unit, even though interpretations of these findings have been controversial. The problem seems to be that primates frequently produce a stream of signs without much evidence for cohesion (e.g., Kanzi, a bonobo, as reported in Savage-Rumbaugh and Lewin 1994). At the same time, it has been reported that Washoe, a chimpanzee who learned how to use signs of American Sign Language, combined the signs for *water* and *bird* to describe a duck (Gardner, Gardner, and Van Cantfort 1989). Kanzi does seem capable of combining a lexigram and a gesture into a meaningful unit (Greenfield and Savage-Rumbaugh 1990).[6]

Interestingly, Kanzi has been reported to have mastered a two-word grammar in his use of lexigrams and gestures, based on the description in Greenfield and Savage-Rumbaugh (1990); see also Heine and Kuteva (2007:145–147). First of all, Kanzi uses two-sign combinations, including creations with one verb and just one argument, in a way that does not distinguish agents/subjects from patients/objects, with both following the verb. While Kanzi's initial combinations (during the first month) show free word order (*hide peanut, peanut hide*), the later combinations seem to converge on the productive verb-noun order (*hide peanut*), even when the noun is the agent, in the sense that the verb is followed by an agent gesture (e.g., *hide Kanzi*). If this characterization of Kanzi's abilities is on the right track, then it is worth pointing out that Kanzi's two-word (verb-noun) grammar is not unlike the grammar behind the verb-noun compounds discussed above.[7] I think this is a good example of how a very specific theory or hypothesis can lead to a clearer revelation of continuity.

However, Washoe's and Kanzi's ability to combine two elements into a meaningful unit should not be taken to mean that they have used compounds or sentences in the same productive and streamlined way that humans do today. The use of such combinations by nonhumans is typically rare and sporadic. The relevant question here is not whether Washoe or Kanzi reached a two-word stage of language but rather whether our common ancestors were in principle capable of (sporadically) combining or interpreting two combined signs. This kind of basic ability, if it was there in our common ancestor, would have greatly facilitated the transition from the postulated one-word (nonsyntactic) stage to the two-word (protosyntactic) stage.

In order for the selection processes to get off the ground, at least some of our ancestors needed to be capable of producing and understanding such combinations. Those who were just a bit better at it would have been the ones whose genes were passed on in the line of descent leading to humans, generation after generation. The speed of the spread depends on how high the fitness of these individuals was relative to the competitors. According to Stone and Lurquin (2007), if relative fitness is high, the increase of the variant in the population can take just a few dozen generations for the variant frequency to increase tenfold.[8]

Based on the proposal outlined above, we can expect to find the ability to combine two signs into a meaningful unit in at least some primates. This is now a testable hypothesis, which stems from a specific proposal about what the initial stage of grammar was like. Moreover, the prediction is that this ability in other primates should be variable, in the sense that some individuals should have a better ability than others, based on their genetic make-up. If that kind of variability was not present in the common ancestor, then there would have been no genetic material for natural/sexual selection to target. In other words, three testable hypotheses can be identified: (i) other primates should in principle be capable of some rudimentary two-sign combinations; (ii) this ability in other primates should be quite variable; and (iii) this ability in other primates is not expected to be identical to the ability in humans.

In this respect, consider Yang's (2013) study, which compares children's combinations of articles (*a* and *the*) and nouns, with the sign combinations by nonhuman primates, of the kind *give X*, or *more X*. He uses a statistical method to demonstrate that children's combinations are consistent with their using a productive rule of grammar, while Nim Chimpsky's combinations do "not show the expected productivity of a rule-based grammar" (1). This is an important study, as it shows that quantitative methods of this kind can be used to probe questions about the origins of human language. However, if my reasoning here is on the right track, then Nim Chimpsky's combinations are not expected to show human-like fluency but a precursor to it, which is what this study seems to have found.[9]

In summary, continuity can be expected only with the most rudimentary of syntactic structures, and what counts as such a structure should be reconstructed based on a linguistic theory, rather than impressionistically. But even there, continuity does not imply that the other primates should show human-like fluency with two-word combinations—not at all. After all, humans had millions of years to undergo selection for language abilities since the common ancestor with, for example, chimpanzees. All one expects or predicts to find in this respect is a clumsy precursor to the ability to combine signs.

There are some amazing findings about the cognition of primates, many of which rely on ingenious experiments, including Seyfarth and Cheney's findings about baboon cognitive abilities. What is needed at this point is for linguists to meet these researchers (at least) halfway by hypothesizing the stages of language evolution, and by arriving at hypotheses that are just as specific and testable as those explored by Seyfarth and Cheney. Without specific hypotheses about language evolution, one cannot draw definitive conclusions about continuity with other primates. Perhaps one can draw a very broad conclusion that these abilities in baboons and other primates constituted a preadaptation for human syntax, in some vague way. But all these abilities are distinct from what we consider to be human language. The essence of language is that it can express or externalize, with some precision, and with combinable and recombinable pieces, various kinds of wishes and thoughts.

My argument here has been that potential points of continuity will best be revealed if one considers the simplest stages in the evolution of human language, rather than the more innovative stages. In order to arrive at such simplest stages, one needs to rely on a linguistic theory for reconstruction. Specific theories reveal specific points of contact/continuity between human language and communication systems of other primates, making it possible to advance specific and testable hypotheses. In addition to looking for language-like abilities in nonhuman primates, continuity should be sought by looking for primate-like abilities in human language.

3

FLUENCY EFFECTS IN HUMAN LANGUAGE

Jennifer E. Arnold

Seyfarth and Cheney argue in "The Social Origins of Language" that the hallmark of human language is the use of a discrete, rule-governed system to achieve social communicative goals. People generally believe that when they talk, they do so intentionally. That is, people say things because they mean them, and other people use those words and sentences to infer the speaker's meaning. Seyfarth and Cheney go on to argue that nonhuman communicative systems share some of these properties, supporting the view that both communicative systems evolved out of similar social systems. In a nutshell, they contend that baboon communication is an intentional and socially functional system. This challenges the perception that animal communication differs categorically from human language, and points toward evolutionary continuity. This argument also bridges the contrast between highly flexible and productive human language systems and relatively inflexible and innate animal communication systems (Cheney and Seyfarth 1997).

I propose here that the comparison between human and nonhuman communicative systems also requires a closer look at human language. In particular, we need to consider the ways in which the intentional aspects of language are rooted in less intentional, automatic aspects of cognition and behavior. I show that unintentional behavior plays a systematic role in the types of human signals that are used communicatively and intentionally. Here I examine this pattern with respect to the ways in which speech variation reflects the information status of the utterance.

When people speak, they exhibit striking variability in the pronunciation of their words. For example, the same speaker might produce the sentence "This food is great" by emphasizing "food" on one occasion—"This *food* is great"—but emphasizing the determiner on another—"*This* food is great." This illustrates one of the most ubiquitous domains of linguistic variation, which is the degree of intelligibility, or acoustic prominence, of a word. This kind of variation contributes to a dimension of language known as *prosody*, which refers to the timing, pitch, rhythm, and acoustic properties of speech. Words vary in acoustic properties such as duration (longer vs. shorter), pitch (e.g., high vs. low, or rising vs. falling), and intensity (loud vs. quiet). Speakers also vary in how clearly they pronounce their words, sometimes attenuating or dropping certain phonemes (*Imuna* vs. *I'm going to*), and varying the degree to which their vowels are distinguishable (Bradlow, Torretta, and Pisoni 1996). In all these dimensions, pronunciations can vary from more *reduced* expressions (short, quiet, unintelligible, low pitch or little pitch movement) to more *prominent* expressions (long, loud, intelligible, higher pitch or more pitch movement).

This variation is interesting because it relates in systematic ways to the message that the speaker is trying to communicate. In particular, it tends to reflect the speaker's assumptions about what the addressee knows or is attending to, or the recent conversational context. This is known as information status.

INFORMATION-STATUS EFFECTS ON ACOUSTIC PROMINENCE VERSUS ACOUSTIC REDUCTION

Imagine that Elise says to Jason, "The new lab computer isn't working." At this point, she can assume that by mentioning the new computer, she and Jason can both assume that they share knowledge of the computer, and additionally are currently attending to it. Previously evoked information like this is termed *given* (or *old*) information, and contrasts with *new* information. Given that Elise introduced the computer into the conversation, they can probably also assume that it will be important in the upcoming conversation, making further mention of the computer relatively predictable. Both givenness and predictability are components of the computer's information status in the discourse (e.g., Chafe 1976; Prince 1981).

There is substantial evidence that information status guides both the way we formulate our ideas in words and the way we understand what other people mean. For example, it sounds more natural to say, "That's my dog. My dog chased the cat" than "That's my dog. The cat was chased by my dog." This demonstrates that speakers are more likely to start their sentence with given information than with new information (Arnold et al. 2000; Bock and Irwin 1980). Listeners also assume that words that come later in the utterance refer to something that is discourse-new (Arnold and Lao 2008).

Information status also guides acoustic prominence versus acoustic reduction. For example, if Elise says, "Where is Sandy?" Jason is likely to then pronounce "Sandy" with an unaccented, acoustically reduced expression, which reflects her status as given and attended. This reflects one of the most reliable effects in acoustic prominence: repeated mention. If the same person or object is mentioned twice in a conversation, the second mention tends to be shorter in duration, lower in pitch, with less pitch movement, and quieter (Fowler and Housum 1987). However, simple remention is not enough. For example, Terken and Hirschberg (1994) asked speakers to describe object movements like "The ball touches the cone, the cone touches the ball." The words *cone* and *ball* in the second sentence were not reduced, unless they appeared in the same roles as they did in the first sentence (as for the second *ball* in "The ball touches the cone, the ball touches the square"). This effect of parallelism may be related to the fact that parallel mention is more expected than nonparallel mention of given information (Arnold 1998), suggesting that reduction is related to the continuity of the discourse. Relatedly, Fowler (1988) has claimed that reduction does not occur unless the word refers to the same referent.

Speakers also tend to use acoustic reduction for information that is predictable from the context. In a classic experiment, Lieberman (1963) demonstrated that speakers pronounced the word *nine* with greater emphasis in an unconstraining context ("The next number you will hear is *nine*") compared to a context that made the word predictable ("A stitch in time saves *nine*"). More recent work has shown that in running speech, words tend to be shorter and more unintelligible if they are statistically likely to co-occur with other words in their context (Bell et al. 2009; Jurafsky et al. 2001), or if they are in a syntactically probable construction (Gahl and Garnsey 2004).

While reduced forms are used for given and predictable information, speakers use fuller, accented forms for information that is new or less predictable. Likewise, prominent (e.g., emphatic-sounding) pronunciations are used when the speaker wishes to communicate contrast. For example, Jason might say, "SANDY is in the lab, but I don't know where KATHRYN is" (Ito and Speer 2008).

IS ACOUSTIC PROMINENCE PRODUCED INTENTIONALLY?

Information-status effects on acoustic prominence are ubiquitous and well established. Yet there is less consensus about why these effects occur. Understanding why is relevant to questions about the continuity of language evolution and, more broadly, the functional nature of communication systems. Seyfarth and Cheney (citing Clark 1996) point out that human language serves specific functions, for example, in that it is typically used for specific social purposes, and always involves both speaker's meaning and addressee's understanding. This view fits well with the traditional view on information-status effects. For example, Grice (1975) proposed that successful human communication depends on speakers and listeners following certain maxims, and assuming that they can interpret others' communications in light of these maxims. One such maxim considers quantity, and suggests that speakers should say just as much as is needed for communication, but not too much. In keeping with this, speakers tend to provide an explicit bottom-up signal in the form of greater acoustic prominence (which confers greater intelligibility) when information is not recoverable from the context—that is, when it is new or unpredictable.

This view represents a functional (or grammatical) explanation of information-status effects. This type of mechanism serves to meet the social/communicative function of language, and it assumes that the acoustic form is chosen because it corresponds to the speaker's intended meaning. In this sense, it is a part of the intentional purposes of speaking, even though speakers do not need to be aware of the mechanisms by which linguistic forms are selected.

The functional explanation predicts that listeners should be able to use acoustic variation to help them identify the right referent, and indeed evidence suggests that they can (Arnold 2008; Dahan, Tanenhaus, and Chambers 2002). For example, Arnold (2008) tracked eye gaze of both adults and children as they followed instructions to move pictures on a computer screen. Subjects viewed a set of four pictures, including two with similar-sounding names (e.g., bagel, bacon). They heard an instruction mentioning one of them: "Put the bacon on the square." The second sentence was the critical one, mentioning the target with either an accented or unaccented expression: "Now put the BACON on the triangle." Critically, at the onset of the target expression "Ba- . . . ," the input was ambiguous, as it could potentially be the start of either *bacon* or *bagel*. When the word was unaccented, listeners had a tendency to look at the unmentioned (discourse-given) object, but when it was accented, this bias disappeared. Thus, prosody helped direct the listener to the more contextually likely referent.

The functional explanation also accords with the fact that people have strong intuitions about the "right" way to say something. For example, it just sounds "wrong" to repeatedly emphasize a referent in a story: "ALEX went to the museum, and ALEX bought a ticket, and ALEX saw the show."

However, the fact that acoustic prominence plays a useful role in communication does not mean that it was produced intentionally. Seyfarth and Cheney (1992) discuss the question of how we (as scientists) can know whether an animal's call is produced intentionally or not. They tell the story of the tennis player Jimmy Connors, who was in the habit of grunting loudly as he hit the tennis ball. When officials complained, he claimed that the grunt was unintentional, just a side effect of physical effort. Without any way to assess this claim, they could not fault him for it.

Similarly, it is possible that the cognitive mechanism that leads speakers to produce variation in acoustic prominence is not directly influenced by the social goal of communication. An alternate possibility is that acoustic variation results from relatively automatic mechanisms, but in such a way that it correlates systematically with information status. If so, perhaps listeners simply learn to make use of acoustic variation as a correlational cue.

Indeed, we know that humans (and other animals) are extraordinarily good at picking up on patterns in their environment, and can use them to draw inferences and make predictions. One example of this comes from evidence that listeners can even use speech disfluency to help them anticipate the speaker's meaning. It turns out that speakers are more likely to be disfluent when they are referring to something new than something given, consistent with the idea that new references require more planning and are cognitively more difficult (Arnold and Tanenhaus 2011). It is implausible that disfluency is produced intentionally as a signal about an upcoming reference to something new.[1] Yet comprehenders can still make use of the systematic relationship between disfluency and discourse newness. In one study (Arnold, Tanenhaus, et al. 2004), we demonstrated that disfluency led to a bias toward discourse-new objects. For example, following "Put the grapes below the candle" (which made the candle "given"), listeners expected the next sentence to include a reference to the candle if the speech was fluent. However, if speech was disfluent, they expected reference to an unmentioned object. Other work has suggested that disfluency leads to a general bias toward things that are difficult to name (Arnold et al. 2007; Heller et al. 2014).

Thus, we must consider the possibility that acoustic prominence may result from other, more automatic constraints on the language production process, and its communicative function derives from the listener's ability to make use of any cues that are available. Indeed, there is extensive evidence that acoustic variation is related to the ease and fluency with which speakers can produce their utterances.

ACOUSTIC PROMINENCE IS RELATED TO SPEECH PRODUCTION FLUENCY

Human language is a complex system, and scholars agree that turning thoughts into words requires manipulating information at numerous levels. For example, speakers need to generate the conceptual structure, think of the words and syntactic structures they need, generate the phonological form for the utterance, and program the motor movements needed to articulate it (e.g., Levelt 1989). Critically, each of these processes takes time and cognitive resources. Thus, anything that changes the ease of planning an utterance might change the speed and fluency with which speakers can produce it. The critical question for our current purposes is whether this changes the prosodic form of a word.

Evidence suggests that planning does indeed affect reference form. Support for this idea comes primarily from evidence that the duration of a word is related to various measures of word planning difficulty (Arnold and Watson 2015; Gillespie 2011; Kahn and Arnold 2012; Zerkle, Rosa, and Arnold 2017). This makes sense, in that planning difficulty likely changes the timing of both planning and articulation—thus, timing variation affects the timing component of prosody, duration.

One effect of planning emerges from the observation that speech planning needs to precede actual articulation. However, speakers have a choice: they can either completely preplan an utterance before saying it or plan as they go along. The on-the-fly option still requires preplanning, but it means that the speaker plans word x while uttering words x-1 and earlier. As a result, difficult words tend to result in slowing on the *previous* word.

Christodoulou (2012) asked participants to name pairs of pictures, for example, "Skunk hand" or "Skunk deer." The critical manipulation was the frequency of the second word, which was either high (*hand*) or relatively low (*deer*). Frequency is well known to influence the speed with which speakers retrieve words, and thus represents a manipulation of relative planning difficulty. Subjects consistently produced a slower "Skunk" preceding a low-frequency word than preceding a high-frequency word. In a series of experiments using eyetracking, Christodoulou critically provided support for the role of utterance planning in this finding. He found that the timing with which participants fixated the second picture (hand or deer) mattered, such that earlier fixations led to shorter first-word durations. The first fixation on the second picture was taken to be a measure of when they began planning the second word (including conceptual processing), suggesting that the duration of word 1 is sensitive to the speaker's degree of readiness to produce word 2. In experiments 3 and 4, results critically demonstrated that the timing of planning word 2 modulated the effect of its duration on word 1. If speakers looked at object 2 *before* they began uttering word 1, there was a strong effect of word 2 frequency, such that word 1 was shorter if word 2 was frequent. However, if speakers began uttering word 1 before fixating on 2, they used a uniformly long duration, regardless of word 2 frequency.

Christodoulou's work illustrates how speech planning can influence the timing of the speech regions that precede the planned unit. Other work shows that ease of planning a speech unit affects the pronunciation of that unit itself. One well-established effect is that words tend to be shorter if they are more probable. A simple measure of probability is frequency, which is a kind of context-free probability—that is, the likelihood of a word in the language overall. Zipf (1929) established a classic effect by which frequent words (e.g., *car*) tend to be shorter than infrequent words (e.g., *automobile*; see also Piantadosi, Tily, and Gibson 2011). This finding extends to the pronunciation of a particular token. For example, homophones like *time* and *thyme* tend to differ in their duration, such that the less frequent member tends to be pronounced with longer durations (Gahl 2008).

Similarly, word duration is sensitive to the probability of a word, contingent on both the preceding and following words. That is, the duration of word *n* depends on how predictable it is based on knowing word *n+1*, and also on how predictable it is based on knowing word *n-1*. A good example of this comes from names (Christodoulou 2012): the word *Ford* is relatively probable following *Harrison* (forward predictability), while the word *Burt* is relatively probable when it precedes *Reynolds* (backward predictability). Bell et al. (2009) found that both preceding and following context matter, such that higher probability words tend to be pronounced more quickly, and with less clarity.

While researchers disagree about the origin of these probability effects, a plausible explanation is that word probability facilitates production: easy-to-produce words are retrieved more quickly, allowing the speaker to plan the following word and produce the two as a unit (Arnold and Watson 2015). This results in fluent, connected speech.

Reduction effects extend beyond just the predictability of words, and also include the predictability of actions. Watson, Arnold, and Tanenhaus (2008) demonstrated this by having pairs of participants play a verbal game of Tic Tac Toe. They informed each other of their moves, saying things like "My marker goes on nine," where the number denoted the grid location. If the move was a winning move, the location was completely predictable to both players, owing to the fact that the goal of the game encourages players to put their marker in a space next to two other markers, if possible. Likewise, a block move was highly predictable. We found that speakers pronounced the number word (e.g., *nine*) more quickly when the move was predictable (i.e., when it was a win/block) than when it was not. In this task, the goal of winning made a move predictable, possibly enabling subjects to begin planning their utterance earlier, pre-preparing the sounds in order to meet fluency. It may also have decreased uncertainty about the move, such that they did not need to slow down to monitor its correctness.

Further evidence that speech difficulty affects prosody comes from analyses of the conditions that lead to disfluency. Speakers often fail to achieve the goal of fluent speech delivery: they hesitate, repeat their words, restart their sentence, or produce filler words like *uh* or *um*. Bell et al. (2003) did a large-scale analysis of the pronunciation of function words like *and*, *the*, and *of* in a speech corpus, and found that words immediately preceding and following disfluent elements tended to have fuller pronunciations. This emerged both in the duration of the pronunciation and in the speaker's choice of the full vowel form (e.g., "thiy" [rhyming with *tree*] vs.

"thuh" for *the*). This suggests that speech difficulty is associated with nonreduced pronunciations.

One proposal for why fluency affects prosody focuses on the sound representations for each word. Watson, Buxó-Lugo, and Simmons (2015) propose that there is a relationship between the ease of retrieving the phonological code for a word and the time it takes to pronounce it. Most models propose that word production involves separate stages for the selection of the word and the selection of the phonological code (e.g., Dell 1986). Watson, Buxó-Lugo, and Simmons argue that speakers may begin to articulate a word before they have finished selecting the phonological code for the end of the word. If so, lengthening the pronunciation would provide the time needed to encode the sound structure. They present data from both human participants and a computational model to support this proposal.

Watson, Buxó-Lugo, and Simmons's proposal is also consistent with evidence that predictable words are shortened. When a word is predictable, the speaker can begin planning it earlier. This would result in the earlier selection of all aspects of the word, including its phonological code.

A related proposal is that acoustic reduction occurs when the word is pre-activated (Kahn and Arnold 2012). Under this view, reduction is more likely when the speaker is already thinking about both the concept and the linguistic word itself, compared with just the concept. Kahn and Arnold demonstrated exactly this pattern, using a task where speakers described moving objects (e.g., "The accordion rotates"). When the accordion was predictable (based on a pre-speech cue), speakers produced the word more quickly. When speakers also heard the actual word *accordion* before their response, the duration was even shorter.

Another possibility is that highly frequent words are easier to produce because the articulatory processes are routinized (Bybee and Hopper 2001). This means that highly practiced sequences are produced more quickly, so it accounts for effects that are learned over time, like frequency or co-occurrence probabilities. However, a variation on this would be needed to account for facilitation effects that arise from the current situation, such as repeated-mention effects.

Another possibility is that planning difficulty of one word can impact the timecourse of planning subsequent words. Consider the sentence "The Venetian vase goes next to the window." If the speaker has momentary difficulty retrieving the word *Venetian* the planning system will require the support of all cognitive resources to successfully retrieve it. This leaves few resources available to concurrently plan the following word, *vase*. This requires the speaker to slow down on *Venetian* while planning the next word. This account predicts that facilitation effects are most likely to occur in the context of multiword utterances, and in a task that encourages on-the-fly planning, as opposed to preplanned speech.

Critically, the fluency effects reported here are somewhat independent of the social/communicative function of language. In one sense they are not completely dissociated, because speech difficulty only occurs when the speaker is attempting to generate an utterance, and this activity is directed by the need to satisfy a social goal. However, fluency effects themselves are not driven by communicative goals. The speaker might intend to introduce a new concept "Look at the Venetian vase," but this creates difficulty, which is the direct cause of duration variation, and perhaps disfluency ("Look at theee uh Venetian vase").

Consistent with this, several studies have found evidence that the intelligibility and duration of spoken words is unaffected by what the listener knows. These findings contradict a hypothesis that acoustic variation is driven by audience design—that is, speakers produce reduced forms precisely when comprehension is facilitated, for example, because the word is repeated or predictable (Lindblom 1990). Instead, many studies have found that variation in the listener's knowledge—for example, whether the listener heard the first mention of the word—has little to no impact on the speaker's tendency to reduce a second mention of the word (Bard and Aylett 2004; Bard et al. 2000; Kahn and Arnold 2015). However, feedback from the listener can affect fluency of utterance planning and production (Arnold, Kahn, and Pancani 2012).

In sum, extensive evidence suggests that the variation in speech pronunciation is systematically related to the process of production. When a word or phrase is difficult to plan, speakers slow down, and become less fluent. They also use phonologically more explicit forms, like full vowels, and resist simplifying consonant clusters. This suggests that the psychological mechanisms of speech production have a direct impact on linguistic form.

WHY FLUENCY EFFECTS MATTER: FLUENCY AND INFORMATION STATUS

The premise here is that it is worthwhile considering the degree to which human language involves relatively automatic mechanisms that are not directly managed by the speaker's social/communicative goals. Above, I demonstrated that the ease of planning an utterance has consequences for linguistic form. These planning effects are most naturally viewed as a side effect of constructing an utterance, suggesting that processing difficulty itself modulates linguistic form.

Yet such effects are not surprising by themselves. What is more interesting is that processing difficulty and speech fluency are systematically related to the kinds of information-status categories that have been proposed to underlie linguistic form for functional reasons.

For example, consider the contrast between "given" (previously mentioned or otherwise evoked) and "new" information. New information is likely to be more difficult to talk about than given information, because it requires more cognitive resources to build a mental representation of the concept, and the words should be harder to retrieve.

Speakers may also allocate more cognitive resources to monitoring words that represent new information, perhaps to check the correctness of the word or message. Moreover, speakers may need more time to construct the phonological representation for new words. This predicts that speakers should slow down for new words, both before and during the word. Indeed, this pattern of reduction for repeated words has been widely reported (Bell et al. 2009; Brown 1983; Fowler and Housum 1987; Halliday 1967).

Yet the processing-based explanation contrasts with the traditional explanation for acoustic variation in spoken language. It is widely accepted that some languages (like English) mark information status via representations in the domain of prosody (e.g., Pierrehumbert and Hirschberg 1990). One aspect of prosody is the placement of a pitch accent, which corresponds to greater acoustic prominence of the word. This grammatical approach can be classified as a part of the functional approach described above, in that it suggests that particular prosodic forms are licensed by grammatical or pragmatic rules, and in particular that prosody is used to mark information status. That is, prosody is a tool for achieving a communicative goal.

For example, Schwarzschild (1999) argues that the grammar includes the constraint that speakers should avoid accenting given information. Thus, a listener can infer that an unaccented word is given, while an accented word is probably new (for other grammatical approaches, see Gussenhoven 1983; Selkirk 1996). While these theories do not specify the mechanisms by which grammar shapes linguistic form, it seems plausible that they should do so directly, such that a particular intended message leads to the selection of the appropriate linguistic form (Arnold 2016; Arnold and Watson 2015; Kahn and Arnold 2012).

Thus, it is notable that information status can have two different types of influences on reference form. The functional (aka "intentional") influence is through the grammar: reduced forms are appropriate in some contexts, such as when the referent is given. The processing (aka "automatic") influence results from the relative ease of production processing that occurs from given or salient information, which leads to shorter and more attenuated pronunciations. Critically, both types of mechanisms have similar effects.

While the correlation between functional and processing accounts is a challenge for researchers, there is reason to believe that processing effects are intertwined with the functional use of prosody. First, timing variation is part and parcel of the accenting categories that are relevant to grammatical theories of prosody. When a word is accented, it affects several acoustic properties, including pitch, pitch movement, intensity, and the relative duration of the word with respect to other words in the utterance (Ladd 1996). For example, Breen et al. (2010) examined the pronunciation of sentences like "Damon fried an omelet" and manipulated the location of focus between Damon, fried, and omelet. They found that focus led speakers to produce words with longer durations, larger f0 excursions, greater intensity, and longer subsequent pauses, compared to when it was not focused. Thus, duration contributes to the expression of information status. Since duration is the dimension of prosody that is most likely to be affected by processing difficulty, this opens the door for speech difficulty itself to affect acoustic prominence.

Moreover, there is evidence that duration variation can influence speech comprehension as well. Using the same experimental paradigm as Arnold (2008), Arnold, Pancani, and Rosa (2015) tracked listeners' eye movements as they followed instructions like "Put the bagel on the square. Now put the bacon on the circle," and compared acoustically prominent words "Now put the BACON . . ." with reduced ones "Now put the bacon . . ." Critically, we also manipulated the overall fluency and speech rate. Half the subjects were told that the speaker was distracted while speaking, and all of the sentences were spoken in a slow and halting manner. This had the effect of slowing all the words in the sentence, including the target word. The question we asked was whether the overall slowing would change the perception of the target word. If only relative prominence matters (as the grammatical approach would suggest), then it should not. By contrast, we found that in the distracted condition, listeners had a stronger bias toward the discourse-new referent than in the undistracted condition. This occurred despite the fact that the slowing was clearly attributable to the speaker's distracted, disfluent delivery, and not the information status itself. On the other hand, listeners were still able to use pitch variation to distinguish the reduced and unreduced conditions, even in the distracted condition. This suggests that acoustic prominence effects cannot be explained entirely in terms of duration variation. Rather, the listener's perception of acoustic prominence stems from all sources of information, including slowing that stems from speaker difficulty.

The research reviewed here focuses on the fact that human speech involves variation in pronunciation. Sometimes speakers use acoustically prominent pronunciations, which are longer, louder, and display greater pitch movement. Sometimes speakers use acoustically reduced pronunciations, which are quieter, softer, and often phonologically simpler.

The classic explanation for this variation focuses on the communicative goal of language: speakers use particular forms to signal the information status of their intended message. Reduced forms generally reflect given or unfocused information, while prominent forms are used for new or unpredictable information. This explanation matches the view that human language (unlike animal communication systems) is characterized as a complex, rule-governed system of communication.

There is no question that languages include grammatical constraints on prosody. Yet in contrast with the traditional view, I have argued for a more complex system. Acoustic variation is not just the result of prosodic marking; it can also result as a side effect of the cognitive processes necessary to produce an utterance. Thus, a reduced pronunciation may not be selected entirely on the basis of the speaker's communicative intention. Nevertheless, processing-based variation impacts the listener's interpretation. Thus, fluency and processing effects are an integral part of prosody.

This view raises new possibilities for identifying continuities between human and animal language communication systems. Seyfarth and Cheney argue that nonhuman primate communication serves many of the same social goals as human language. On the flip side, I argue that human language incorporates some relatively automatic processes—a type of mechanism that is often attributed to animal systems. Fluency effects may not be designed specifically for a social purpose, but they correlate with more purposeful grammatical effects. Thus, fluency contributes to successful communication, the ultimate goal of language use.

*   *   *

This work was partly supported by NSF grant 0745627 to J. Arnold.

4

RELATIONAL KNOWLEDGE AND THE ORIGINS OF LANGUAGE

BENJAMIN WILSON AND
CHRISTOPHER I. PETKOV

The question of how human language evolved is a mesmerizing and perplexing puzzle with many of the pieces missing. The fossil record provides limited insights into the neurolinguistic capacities of our recent ancestors. Often we must infer how the language system is likely to have evolved from the evidence that can be obtained in nature or in the laboratory from studying the behavior and neurobiology of extant animals.

Some approaches to understanding how language evolved focus on the differences between humans and nonhuman primates. For instance, in terms of vocal production, nonhuman primates do not appear to combine their vocalizations into structured sequences, which humans, and by convergent evolution, songbirds and a select few other species readily do (Egnor and Hauser 2004; Berwick et al. 2012; Bickerton and Szathmary 2009; Hurford 2012). The relatively rudimentary vocal production abilities of many mammals, nonhuman primates included, appear to coincide with differences in the pathways for vocal motor control in relation to those in humans and songbirds (Simonyan and Horwitz 2011; Jurgens 2002; Petkov and Jarvis 2012; Chakraborty et al. 2015). Other perspectives have emphasized the similarities between humans and other animals (Seyfarth and Cheney

1999; Seyfarth, Cheney, and Marler 1980; Jarvis 2004; Hurford 2007), such as the ability to form complex social networks (Cheney, Seyfarth, and Smuts 1986; Bergman et al. 2003; Dunbar and Shultz 2007). A third vantage point is to understand whether more closely evolutionarily related species show gradations in certain abilities relative to more distantly related species. This could help to clarify the path toward vocal production learning and ultimately human language by deconstructing the core behavioral phenotypes and neurobiological substrates that changed in form or capacity during language evolution (Chakraborty et al. 2015; Feenders et al. 2008; Petkov and Wilson 2012). We expect the origins of language to be informed by all three approaches: similarity, differences, and gradations in capacities.

It has also been useful to distinguish between vocal production learning and auditory receptive learning (Jarvis 2004; Petkov and Jarvis 2012). This distinction is based on the notion that vocal production and receptive learning abilities engage distinctly different brain processes and pathways. Such a perspective allows us to look beyond the vocal production abilities of many mammals and to consider species that might otherwise be easily dismissed because of their seemingly relatively simple vocal production capacities. We might also predict and test for gradations in abilities and neurobiological substrates that support them, which would not be as accurately informed by a categorical behavioral distinction: a presence or absence of a given ability (Petkov and Jarvis 2012; Arriaga and Jarvis 2013; Feenders et al. 2008). To explore the receptive learning capacities of different animals requires assessing animal learning using paradigms that allow manipulating different levels of complexity in perceptual learning alongside measures of behavioral responses that allow investigators to decode what the animals can learn and the learning strategies that they adopt (Wilson, Marslen-Wilson, and Petkov 2017).

A key feature of human language syntax is the ability to evaluate the grammatical relationships between words in a sentence (Bickerton and Szathmary 2009; Fitch 2010). How did such an impressive combinatorial capacity evolve? One candidate precursor to language is the general capacity to learn relationships between sensory elements in a sequence (Fitch and Friederici 2012). As we will consider, convergent evidence from a range of scientific fields suggests that a number of nonhuman animals are able to process certain relationships between items or elements in sequences of auditory or visual stimuli. This is being studied across a range of tasks and paradigms in a number of nonhuman primate species. We briefly overview these tasks here and consider them in greater detail below.

Sequence processing tasks and Artificial Grammar Learning (AGL) paradigms have been used to emulate aspects of language syntax, such as certain relationships between words in a sentence (Reber 1967). Such tasks have been used to investigate preverbal infants' language-related learning abilities (Marcus et al. 1999; Saffran et al. 2008). Such nonlinguistic paradigms are also well suited for studying nonhuman animals. AGL tasks have been used to show that a number of animals have impressive capacities to learn certain relationships between sounds or pictures in a sequence (Fitch and Hauser 2004; Hauser and Glynn 2009; Gentner et al. 2006; van Heijningen et al. 2009; Stobbe et al. 2012; Sonnweber, Ravignani, and Fitch 2015; Ravignani et al. 2013; Newport et al. 2004; Wilson, Smith, and Petkov 2015; Wilson et al. 2013; Murphy, Mondragon, and Murphy 2008).

Generally in AGL experiments, the AG will be based on one or more rules regulating how the elements in a sequence are organized, establishing the ordering relationships between the constituent elements. Participants are exposed to a subset of the legal sequences that the AG can generate, from which they can extract regularities in the ordering relationships between the sequence elements. Then the participants are tested with novel legal sequences that follow the AG ordering relationships and sequences that violate the rules or relationships in specific ways. AGs can also implement ordering relationships of different levels of complexity. Sequences may only have dependencies between adjacent elements, which infants seem to learn at an early age (Gervain et al. 2008; Marcus et al. 1999), and/or nonadjacent dependencies between temporally more distantly separated elements in the sequences, which human infants appear to learn in the first year of life (Saffran et al. 2007; Gómez 2002; Friederici 2011). Some AGs incorporate complex hierarchical relationships, which might be difficult for nonhuman animals and even humans to learn (van Heijningen et al. 2009; Berwick et al. 2011). Although humans can also combine meaningful words in structured sentences, AGs allow scientists to directly study sequencing structure without semantics. Thus, in AG sequences none of the elements are meaningful by themselves; it is the ordering relationships that matter.

A range of rodents, birds, monkeys, apes, and humans show signs of being able to learn relationships at least between elements that are adjacent in a sequence (Reber 1967; Fitch and Hauser 2004; Hauser and Glynn 2009; Gentner et al. 2006; van Heijningen et al. 2009; Stobbe et al. 2012; Sonnweber, Ravignani, and Fitch 2015; Ravignani et al. 2013; Newport et al. 2004; Wilson, Smith, and Petkov 2015; Wilson et al. 2013; Murphy, Mondragon, and Murphy 2008). Furthermore, many species, including nonhuman primates, are able to learn sequencing relationships that appear to be considerably more complex than the sequences of vocalizations that they naturally produce (Arnold and Zuberbuhler 2008).

However, the growing evidence from animal AGL paradigms raises interesting questions. For instance, if monkeys and apes do not use these capabilities in their vocal communication, what function might these sequencing abilities have evolved to support? One interesting possibility is that AGL paradigms are tapping into a generic system that allows the animals to learn the relationship between co-occurring events, including those more distantly separated in time, such as in sequences containing nonadjacent dependencies. This represents a form of Hebbian learning (Hebb 1949), where activity in a set of neurons A co-occurs with depolarization of neurons B, which leads to the strengthening of synaptic connections between the neural substrates that represent the two events and to an association forming between the events separated in time. This can coincide with neurons changing their responses to A and B to be more similar (Messinger et al. 2001), after the association between the two events is established. For a behavioral example, vervet monkeys learn that a particular alarm call produced by a conspecific is predictive of the presence of a certain predator in the vicinity (Seyfarth, Cheney, and Marler 1980). Another example from the same research group is based on insights into the social knowledge capabilities of baboons in field studies in the Okavango Delta in Botswana (Bergman et al. 2003). The investigators showed that when a baboon listened to the sequence of vocalizations produced during an aggressive interaction between two other individuals, the baboon was able to interpret the sequence of vocalizations based on the rank of these two animals in the social dominance hierarchy.

In this chapter, we will consider this interesting example of baboon classification by rank and kinship in greater detail and discuss the potential links that could be made with the insights that have been obtained using AGL paradigms. We propose that a common thread, between the field studies of baboon social knowledge during the perception of vocal exchanges and the forms of sequence processing studied in AGL paradigms, is that they tap into a more general relational knowledge system. This system depends on the capacity to learn combinatorial relationships between important environmental events in different scenarios and as such may provide clues about the origins of language.

We consider the following guiding questions: What are the links between relational knowledge in social cognition and sequence processing in AGL paradigms? What key neurobiological substrates are likely to be involved? Where are the remaining epistemic gaps that could be better bridged with future study?

WHAT ARE THE LINKS BETWEEN RELATIONAL KNOWLEDGE IN SOCIAL EXCHANGES AND SEQUENCE PROCESSING?

Baboon Social Rank and Relational Social Knowledge

The baboons in the Botswana natural reserve that Seyfarth and Cheney have been studying live in large social groups with complex social hierarchies. While the males' social positions are variable (male baboons join and leave groups relatively frequently), female dominance hierarchies are inherited and stable. The dominance hierarchy is organized both by matrilines, where all related females share a broadly comparable rank, and by individuals, where each female

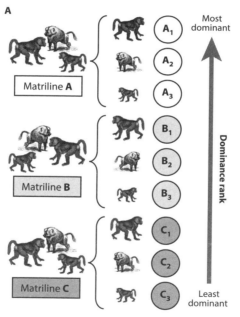

Figure 3 (A). An individual's dominance is determined by their family group, whereby any member of the highest ranking family (Matriline A) will be dominant to all monkeys in lower ranking families. An additional level exists in the dominance hierarchy within each family.

within a matriline will possess her own rank (see fig. 3A and Bergman et al. 2003).

Being social primates, baboons produce specific vocalizations in different social contexts. For example, a "threat-grunt" is a baboon vocalization produced when an animal is being aggressive, while a "scream," along with the associated facial and postural cues, denotes submission (Bergman et al. 2003). Critically, these interactions typically only occur in line with the troop's dominance hierarchy. Specifically, domi-

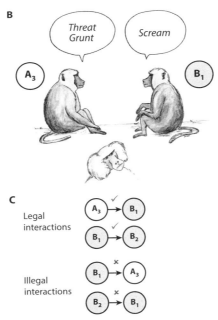

Figure 3. (B) A legal interaction in which a dominant animal threatens a subordinate, who produces an appropriate submissive response. (Artwork by J. Locke.) (C) Such an interaction played back through a hidden audio speaker elicits little response in the listening individual. However, when an illegal interaction is played back instead, it elicits longer orienting responses. The listening animal notices a violation in the expected social interaction, which can only occur if she knows the social rank of the animals interacting in the vocal exchange.

nant animals will take the role of the aggressor while subordinate animals will respond submissively, resulting in an exchange that often avoids physical conflict and injury. The vocal exchanges are predictable if one knows the animals' social rank in the dominance hierarchy. Rarely, a subordinate animal might challenge a dominant animal for social rank and thus produce aggressive calls. A successful challenge will result in a role reversal with social consequences for the animals whose rank changes. However, if a rank reversal occurs, it tends to occur within a family group. A rank reversal between family groups is much more severe because it means that the social position of the two families would switch and all of the animals within a family would change their rank relationships with all of those in the other family. This would happen, for instance, when a female from a subordinate family successfully challenges a member of a dominant family: the subordinate family thereby becomes the dominant family, destabilizing the social rank of many baboons. Most of the time, the baboons know that it is unlikely that a subordinate baboon would produce a threatening vocalization toward a dominant animal, and that the dominant monkey would respond submissively.

Bergman and colleagues conducted playback experiments manipulating the social conditions as follows. An individual baboon (depicted as the baboon listening to the social exchange in fig. 3B) was presented with a prerecorded sequence of vocalizations from a hidden audio speaker (Bergman et al. 2003). The sequence of vocalizations heard would represent either legal or illegal social interactions between two other animals to see if the animal noticed the difference (e.g., fig. 3B-C). A legal vocal sequence would involve an aggressive vocalization produced by a dominant animal followed by a submissive call from a subordinate. Conversely, an illegal, unexpected interaction would involve a subordinate animal vocalizing aggressively, followed by a submissive call from a dominant animal, contrary to the established dominance hierarchy. The listening animals responded notably more strongly (looking longer toward the direction of the hidden speaker) to call sequences associated with illegal social interactions than to legal call sequences. Furthermore, stronger responses were seen when illegal interactions occurred across, rather than within, family groups, in line with the relative severity such a family-wide reversal of social rank within the dominance hierarchy would produce.

Seyfarth and Cheney note in part 1 of this book that experiments such as these show that the animals are able to learn social relationships across different levels (within and between families) and can apply this social knowledge to a vocal exchange reflecting either legal or illegal social interactions. They argue that this form of social knowledge is similar to human social interactions and of notable interest for understanding the origins of human language (Seyfarth and Cheney 2014). We also find these observations from fieldwork interesting, along with the potential links that may exist between how animals evaluate vocal sequences in the wild and how they perform on AGL paradigms in the laboratory. In AGL studies, a number of nonhuman primates show sequence processing capacities that, to some extent, resemble the ability to assess the exchange between the vocalizations produced by different baboons in the wild.

Artificial Grammar Learning Paradigms and Animal Behavior

A number of research groups have taken a different approach toward exploring questions about the evolution of language. These studies focus on the animals' abilities to learn the regularities between elements in a sequence. Such paradigms also often evaluate the level of complexity of sequencing relationships that different animals are able to learn.

In such experiments, a human or nonhuman participant is exposed to a representative set of sequences that follow a certain set of rules. This occurs through implicit exposure where the animal needs only to, in the case of sequences of sounds, listen to the sequences for some length of time. Subsequently, the participant is tested with sequences that either are consistent with the AG rule(s) or violate them in certain ways. Different response measures have been used, including natural orienting responses in field or lab settings and go/no-go or explicit button press responses (for an overview, see Fitch and Friederici 2012). Some AGL studies with nonhuman animals have been conducted via operant training where the animal is rewarded for producing a response (e.g., pressing a button or lever) following legal, but not illegal, sequences (e.g., Gentner et al. 2006; Murphy, Mondragon, and Murphy 2008). In all of these studies, different behavioral responses to the legal and illegal sequences suggest that the participant noticed the violations and something about the regularities instantiated by the AG.

A wide range of AGs has been used to explore animal sequence processing abilities. The AGs used vary in a number of different ways and are not always easy to relate in terms of their complexity. Here, we summarize some of the key features of different AG paradigms and refer the reader to the literature for a more comprehensive treatment of sequencing complexity and quantitative comparisons of different AGs (Wilson et al. 2013; Honda and Okanoya 1999; Hurford 2012; Wilson, Marslen-Wilson, and Petkov 2017).

Some AGs generate patterns consisting of adjacent relationships between stimuli, for example, sequences of three tones of the form AAB or ABA, where A and B represent high and low pitch tones respectively (Murphy, Mondragon, and Murphy 2008). Other studies use AGs to produce more variable (nondeterministic) sequences consisting of both optional and obligatory elements (Reber 1967; Petersson, Folia, and Hagoort 2012; Saffran et al. 2008; Abe and Watanabe 2011; Wilson, Smith, and Petkov 2015; Wilson et al. 2013), which can occur in a wide range of legal orders (see fig. 4 for an example).

Some AGL paradigms have assessed the abilities of nonhuman animals to learn relationships between different classes of stimuli, based on specific perceptual features (Fitch

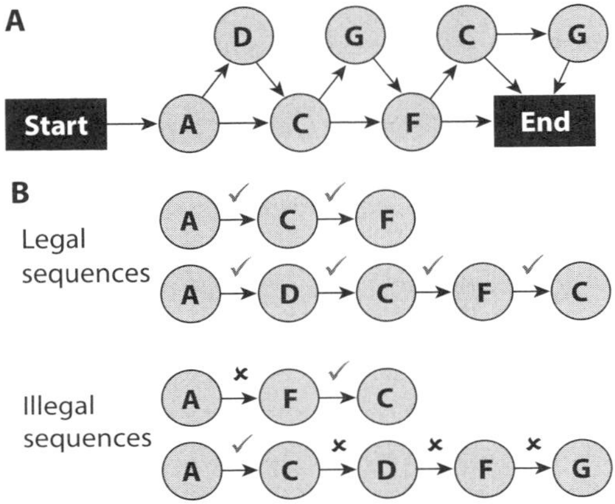

Figure 4. (A) The state transitions of a right branching, non-deterministic Artificial Grammar with several elements (A, C, D, F, G; nonsense words, see Wilson et al., 2013; 2015). The AG can generate sequences of variable length and composition. Following the arrows generates a legal sequence of transitions (check marks over legal transitions in B). Not following the arrows generates an illegal transition (x's over the transitions in the illegal sequences in B).

and Hauser 2004; Hauser and Glynn 2009; Gentner et al. 2006; van Heijningen et al. 2009; Stobbe et al. 2012; Comins and Gentner 2013; Toro and Trobalón 2005). For example, Fitch and Hauser (2004) presented cotton-top tamarins with sequences of nonsense words spoken by either a male or a female speaker. They found that the monkeys were sensitive to violations of a simple, alternating pattern of the form ABABAB (where A and B represent syllables produced by a female and a male speaker respectively), but not to more complex sequences of the form AAABBB. These AGs have been used to test several avian species—starlings (Gentner et al. 2006; Comins and Gentner 2013), Bengalese finches (van Heijningen et al. 2009), pigeons and kea (Stobbe et al. 2012). All of these species have been reported to notice certain violations of both the ABABAB and AAABBB sequences. However, in many cases violations of the AAABBBB sequences appear to be identified using relatively simple learning strategies, such as the animals noticing the perceptual features that identify a certain number of As followed by a certain number of Bs (Beckers et al. 2012; Berwick et al. 2011; van Heijningen et al. 2009).

Some studies have also seen differences in the level of AG sequencing complexity that different species are able to process, which provides some evidence for a gradation in sequence processing abilities. As mentioned above, tamarin monkeys (New World monkeys) and humans can both notice violations to adjacent, alternating AB relationships, while humans seem to be better at detecting more complex relationships in sequences of the form AAABBB (Fitch and Hauser 2004), including when there are explicit ordering relationships between particular A and B elements, as in $A_1 A_2 B_2 B_1$ (Bahlmann, Schubotz, and Friederici 2008; Udden et al. 2012). In other behavioral experiments, both rhesus macaques (Old World monkeys) and common marmosets (New World monkeys) notice certain violations of ordering relationships in sound sequences (Wilson et al. 2013). The marmosets appear to primarily notice violations in just the initial positions in the AG sequences (Wilson et al. 2013). However, the macaques and human participants also notice violations of adjacent relationships throughout these sequences of five to six elements (Wilson, Smith, and Petkov 2015). The AG that was used also contained nonadjacent relationships, which the macaques did not appear to notice and which only some of the humans were sensitive to, revealing both cross-species and within-species differences in sequence processing capabilities. Work with birds has also noted within species (van Heijningen et al. 2009) and across species (Spierings and ten Cate 2016; Stobbe et al. 2012) differences in avian AGL capabilities. Other gradations in animal abilities are being assessed in rodents and birds (e.g., Arriaga and Jarvis 2013; Petkov and Jarvis 2012; Feenders et al. 2008). For example, the impressive vocal imitation abilities of parrots in relation to songbirds appear to be supported by changes in neurobiological substrates. The parrot forebrain song nuclei have a core and shell organization, whereas songbirds appear to have only a core forebrain song nucleus (Chakraborty et al. 2015).

Other AGs have been designed to assess the learning of nonadjacent dependencies. For example, Newport and colleagues (2004) presented cotton-top tamarin monkeys (New World monkeys) with sequences of three syllables in which the first syllable predicted the final syllable, separated by an uninformative middle syllable. In a series of experiments, they showed that under a number of conditions the monkeys were sensitive to these nonadjacent relationships. Also, spider monkeys have been tested with sequences of high and low tones, matched to the animals' naturally vocal frequencies (Ravignani et al. 2013). In this paradigm, legal sequences were required to begin and end with the same tone, separated by a number of tones at a different center frequency (e.g., ABBBA, where A and B represent high and low frequency tones, respectively). The authors found that the monkeys noticed when the final A tone was absent (e.g., ABBBB). Finally, operant conditioning with reward has been used to train chimpanzees to recognize certain nonadjacent relationships. Here, a sequence of visual images were presented simultaneously, with the leftmost image predicting the rightmost image, while the central image(s) provided no useful information (Sonnweber, Ravignani, and Fitch 2015). These studies provide evidence that monkeys and apes are sensitive to relationships between temporally and spatially separated stimuli, in both the auditory and visual modalities, suggesting that the ability to learn these relationships may not be limited to any particular sensory domain. However, it is notable that if adjacent cues are sufficiently informative, then nonadjacent relationships are sometimes not recognized by humans (Gómez 2002) or monkeys (Wilson, Smith, and Petkov 2015).

These AGL studies show that a wide range of species, including New and Old World monkeys, can appreciate the relationships between otherwise arbitrary auditory or visual stimuli and recognize when ordering relationships are illegal, following either exposure or training with exemplary sequences generated by AGs. This capacity includes the processing of adjacent relationships, and some species also show a sensitivity to nonadjacent relationships or sequences with greater levels of variability in their length and composition (Wilson et al. 2013; Fitch and Hauser 2004; Murphy, Mondragon, and Murphy 2008; Gentner et al. 2006; Hauser and Glynn 2009; Friederici et al. 2006; Ravignani et al. 2013; Stobbe et al. 2012; Wilson, Smith, and Petkov 2015).

Similarities and Differences between Field Studies on Social Knowledge and Sequence Processing

How might the abilities required by AGL experiments relate to or differ from those involved in processing vocal social interactions in the baboons? First, we consider the cognitive processes that the Okavango Delta baboons likely need to employ in order to differentiate between a legal and an illegal social interaction. They must have an understanding of the dominance hierarchy of the troop, including the individual relationships between every possible dyad. This is no small task in a social group of up to eighty individuals, but the animals are well motivated to stay vigilant to social rank between and within families. Furthermore, they must understand that certain vocalizations occur only within a specific social context; in an interaction between any two individuals, only the dominant animal instigates the aggressive interaction, and the subordinate produces a submissive response to avoid physical aggression.

On the surface these abilities bear little resemblance to those assessed by AGL paradigms, although on closer inspection there are interesting similarities. All AG tasks test participants on their ability to learn the legal ordering relationships between different stimulus elements or stimulus classes. The processing of sequencing relationships in AGs is thus similar to the baboons evaluating the relationship between the vocalizations produced in sequence between two individuals with known social rank. Of course, the example with the Botswana baboons requires that the sequence of specific, meaningful vocalizations is integrated with social knowledge of the baboon troop's dominance hierarchy, which is learned by interacting with or observing other group members. The AGL paradigms, on the other hand, depend on the acquisition of arbitrary relationships between auditory or visual stimuli, often by implicit learning, but there is no requirement of attaching particular meaning to the elements in the sequence, only that the animal is able to perceive the elements as distinct perceptual objects. Also, AGs can be used to generate much longer sequences often containing adjacent, nonadjacent, and other relationships, whereas in the baboon social knowledge experiments the sequence of vocalizations was short and of considerable potential importance for the animal listening.

These differences notwithstanding, the similarities between the field and laboratory paradigms that we have been considering are intriguing. The social knowledge studies of Bergman and colleagues appear to be similar to AGL studies in some way. First, in these interactions the aggressive threat call always precedes the submissive response. This is a relatively trivial ordering relationship, but nonetheless resembles the adjacent relationship between two elements in a sequence (A predicts B). More interestingly, the dominance rank of the first, aggressive caller (A) predicts the rank of the baboon who produces the submissive response (B). The submissive response is expected only from a subordinate, lower-ranking individual (so $A_{DOM} B_{SUB}$ represents a legal interaction, while $A_{SUB} B_{DOM}$ is highly unexpected and would elicit a much stronger response). Although the baboon vocal exchange is based on the social ranks of the individuals involved, whereas

AG relationships are arbitrary and implicitly learned, both present cases of animals listening to and appreciating the relationships between sounds that occur in expected, "legal" relationships, or in unexpected, "illegal" ones (figs. 3 and 4). Therefore, while different on the surface, both sets of paradigms seem to tap into a relational learning system that has the capacity to apply knowledge of particular forms of relationships and to assess incoming information for consistency or inconsistency with the animal's knowledge.

This is important because in terms of baboon or macaque vocal production abilities it is unclear whether these Old World monkeys combine their vocalizations in sequences, as has been shown for certain species, which produce combinations of two or more vocalizations (e.g., Arnold and Zuberbuhler 2006). Even if future evidence is obtained that baboons and macaques also combine vocalizations, it is interesting nonetheless that the receptive learning capacities of certain nonhuman animals appear to outstrip their vocal production capabilities. Altogether, if the common ground between the two paradigms revolves around relational learning, we might consider that a generic sequence order processing system in the brains of the animals (Frost et al. 2015) interacts with regions that carry social or other information, providing the flexibility to apply relational knowledge under different scenarios, even when events are separated in time.

WHAT NEUROBIOLOGICAL SUBSTRATES MIGHT BE INVOLVED?

We still do not understand the full behavioral receptive learning capacities of any animal species, let alone the abilities of enough species to draw strong evolutionary or phylogenetic conclusions. Yet, as our understanding of behavioral abilities and their neurobiological substrates grows, it is important to reassess our understanding in order to generate new empirical predictions. Work is already underway to understand the neurobiological substrates of adjacent and nonadjacent sequencing processes in songbirds (Lu and Vicario 2014; Abe and Watanabe 2011) and monkeys (Attaheri et al. 2015; Meyer and Olson 2011; Wilson et al. 2015; Milne et al. 2016; Wang et al. 2015; Uhrig, Dehaene, and Jarraya 2014; Wilson, Marslen-Wilson, and Petkov 2017; Dehaene et al. 2015). There is a much larger literature of studies in humans using AGL paradigms of different levels of complexity, which have been used to advance neuroevolutionary hypotheses that can be tested in nonhuman animals (e.g., Friederici 2011; Petersson, Folia, and Hagoort 2012). Independently, our understanding of the neurobiology of social knowledge is also growing and should inform this discussion (Sallet et al. 2011; Platt, Seyfarth, and Cheney 2016). We consider how sequencing processes and social knowledge are likely to engage certain neurobiological substrates in humans and other animals.

Human functional Magnetic Resonance Imaging (fMRI) studies have provided insights into the processing of AG sequences and how this relates to the processing of natural language material. Brain areas involved in natural language processing involve "perisylvian" frontal, temporal, and parietal areas, parts of which can also be activated by AGL paradigms (Petersson, Folia, and Hagoort 2012; Hickok and Poeppel 2007; Friederici 2011). A wide range of natural language tasks engage Broca's territory in the left inferior frontal gyrus, and damage to this region produces language impairments (for a review, see Friederici 2011). More complex AGL paradigms, such as those that involve hierarchical nonadjacent relationships of the forms present in language, can also engage this area (Bahlmann et al. 2009; Bahlmann, Schubotz, and Friederici 2008; Friederici et al. 2006; Petersson, Folia, and Hagoort 2012; Petersson, Forkstam, and Ingvar 2004). Electrical stimulation of this region, during either the learn-

ing or testing phase, can enhance participants' ability to identify AG violation sequences (Udden et al. 2008; de Vries et al.
2010).

Beyond the involvement of Broca's territory in certain
language-specific processes and AGL paradigms, other frontal areas seem to be involved in more general sequence processing functions, some of which appear to be evolutionarily
conserved in function in nonhuman primates (Friederici et al.
2006; Friederici 2004; Wilson et al. 2015). For instance, the
initial stages of human syntactic processing involve evaluating the grammatical relationships within or between adjacent
phrases. This engages the frontal operculum and anterior insula in particular, adjacent to Broca's territory (Friederici and
Kotz 2003; Ni et al. 2000; Friederici, Opitz, and von Cramon
2000; Friederici 2011). When human participants are tested
with AGL paradigms, involving adjacent relationships between alternating elements in sequences of the form ABAB,
this area is strongly engaged (Bahlmann et al. 2009; Bahlmann, Schubotz, and Friederici 2008; Friederici et al. 2006).
Furthermore, although the human frontal operculum and
Broca's territory are adjacent to each other in the frontal
cortex, they appear to be interconnected with different parts
of the temporal lobe by way of distinct white matter tracts.
Namely, Broca's territory is connected to auditory areas in
the temporal lobe by way of a dorsal tract (the arcuate fasciculus) that may have specialized in humans, although how
exactly remains controversial (Romanski et al. 1999;
Bornkessel-Schlesewsky et al. 2015; Friederici 2011; Rauschecker 1998; Neubert et al. 2014; Frey, Mackey, and Petrides 2014; Rilling et al. 2008). The frontal operculum is
part of a ventral pathway that interconnects more anterior
temporal lobe auditory areas with the frontal cortex (Romanski et al. 1999; Bornkessel-Schlesewsky et al. 2015; Friederici 2011; Rauschecker 1998; Petkov et al. 2015).

Therefore, human AGL paradigms have identified at least
two sequence processing pathways that depend on a network

of regions interconnecting sensory cortex to various frontal regions, depending on the complexity of the sequence processing demands. The extent to which these regions or pathways are evolutionarily conserved in function is currently under active investigation (Wilson, Marslen-Wilson, and Petkov 2017). Auditory neuronal recordings in songbirds have provided insights into how neurons in auditory and visual cortex respond to adjacent and nonadjacent sequencing relationships (Lu and Vicario 2014) and neuronal recordings in macaques show how inferior temporal cortex neurons respond to visual sequencing relationships (Meyer and Olson 2011). Two recent studies used deviance detection (oddball) paradigms to assess how the brains of humans and monkeys respond when an unexpected stimulus is heard during a sequence of sounds, or when an expected sound is omitted (Uhrig, Dehaene, and Jarraya 2014; Wang et al. 2015). When the sequence of sounds is predictable (e.g., the same tone is repeated several times), the presentation of an unexpected sound (a tone of a different pitch) produces activation in auditory cortex. However, when more complex sequences are presented (e.g., a sequence of tones following the pattern AAAB, where A and B represent tones of two different pitches), then the absence of the final B element (i.e., in the sequence AAAA) engages a broader set of regions, including bilateral insula and ventrolateral prefrontal cortex in both species (Uhrig, Dehaene, and Jarraya 2014; Wang et al. 2015).

An AGL task, involving more variable sequence ordering relationships, has been used to assess human and monkey sequence processing (Wilson, Smith, and Petkov 2015; Wilson et al. 2013; Saffran et al. 2008). The same paradigm has also recently been used with comparative fMRI in humans and monkeys (Wilson et al. 2015). Following exposure to structured sequences of auditory nonsense words, unexpected "violation" sequences produced highly comparable patterns of activation in the frontal operculum and anterior insula in both humans and monkeys (Wilson et al. 2015), areas known to be involved in processing adjacent relationships in humans (Friederici 2011). These results suggest that the capacity to process adjacent relationships between elements in a sequence, and the brain areas that are involved in human adjacent sequence processing and grammatical operations, share mechanisms with evolutionarily conserved functions in the corresponding frontal operculum regions in extant nonhuman primates (Wilson, Marslen-Wilson, and Petkov 2017).

Concurrently, there is considerable interest in understanding how the brains of humans and nonhuman primates support social cognition (Platt, Seyfarth, and Cheney 2016; Rushworth, Mars, and Sallet 2013). Social network size is known to correlate with the size of various brain areas in humans, including the amygdala (Bickart et al. 2011), STS/STG regions (Kanai et al. 2011), orbitofrontal cortex (Powell et al. 2012), and ventromedial prefrontal cortex (Lewis et al. 2011). These areas are implicated in social tasks, such as recognizing conspecifics or inferring mental states (Adolphs 2009; Frith 2007) or the value of social information (Behrens, Hunt, and Rushworth 2009). In monkeys, social group size predicts gray matter thickness in areas that correspond to those associated with social cognition in humans, such as the STS, inferior temporal regions, rostral prefrontal cortex, temporal pole, and amygdala (Sallet et al. 2011). Moreover, the thickness of prefrontal cortex also scales with rank as well as social group size (Sallet et al. 2011). Many of these areas are implicated in social perception and cognition in both humans and monkeys (Rushworth, Mars, and Sallet 2013; Rudebeck et al. 2006). These studies demonstrate that brain structure and function covary with social network size in both species (Rushworth, Mars, and Sallet 2013), and social information appears to be processed by homologous regions in humans and nonhuman primates (Platt, Seyfarth, and Cheney 2016).

The regions that are sensitive to social content and those that are involved in sequence processing do not tend to overlap. Therefore, it is not clear how the regions involved in sequence processing and social cognition may interact. As we have considered, the baboons' responses to sequences of vocalizations from different individuals requires the processing of social information, likely including engaging voice-sensitive areas in the temporal lobe (Petkov et al. 2008; Belin et al. 2000). Applying knowledge of the functional meaning of the vocalizations as well as the position of the baboon callers within the dominance hierarchy may well engage a number of other brain regions in the temporal and frontal cortex (Sallet et al. 2011). Assessing the order in which the vocalizations occur, we speculate, may involve ventral frontal regions sensitive to sequence ordering relationships. Deconstructing the behavioral phenotypes involved and developing paradigms that allow manipulating social cognition and sequencing relationships is undoubtedly going to be critical for interpreting neurobiological data on joint operations subserving these interesting natural behaviors.

EPISTEMIC GAPS: PATHWAYS FOR FUTURE BEHAVIORAL AND NEUROBIOLOGICAL STUDY

We conclude by summarizing some of the implications of this discussion, which might be of interest for future scientific study. What does a general relational learning and knowledge system predict that might give us new insights into the origins of human language? How could we strengthen the links between observations in the field and laboratory studies on social learning and sequence processing?

We began by suggesting that understanding gradations in abilities could be just as informative as understanding cross-species similarities and differences. The idea that a relational learning system provides the substrate for human combinatorial capacities in language predicts phylogenetic differences in relational learning capacities. For instance, some animals might be able to process longer sequences of sensory events, better manage with informative and noninformative transitional regularities, or process greater levels of complexity in nonadjacent relationships separated by multiple intervening elements. A similar set of predictions might be made of species with exceptional vocal production learning or imitation abilities, such as songbirds, cetaceans, and other animals capable of complex vocal production learning, in relation to species closely evolutionarily related to these animals.

One might also predict and comparatively test for a gradation in social knowledge abilities in different nonhuman primates. Brain size is one feature known to correlate with social group size in primates, and indeed the maintenance of social relationships in increasingly large groups has been hypothesized to be one of the key evolutionary pressures that led to the rapid expansion of the primate brain (Dunbar and Shultz 2007). This does not require focusing solely on species that already have elaborate social hierarchies in the wild and is something that can be studied in laboratory primates that share those capacities for social knowledge (Shepherd and Platt 2010; Sallet et al. 2011; Platt, Seyfarth, and Cheney 2016). We considered initial behavioral evidence from AGL studies in primates that have provided evidence in support of a gradation in sequence processing capabilities in species closely evolutionarily related to humans.

There are also opportunities for AG studies to implement social stimuli (pictures of monkey faces or the use of vocalizations) within sequences to evaluate how the social content influences AG learning. In some scenarios, the AG relationships might be better learned using ethologically meaningful stimuli. Alternatively, social knowledge could be manipulated to disrupt the learning of certain sequencing relationships, such as when the sequencing relationships conflict with social knowledge. Likewise, one could ask questions about whether baboons (or other animals) in the wild could evaluate multiple social interactions in a sequence, whether the sequence ordering relationships matter to the animals, or what the limits of such abilities are during exchanges between multiple individuals.

In closing, we may lack all of the pieces to the language evolution puzzle. However, this problem is as enticing as it is challenging, precisely because with many of the pieces missing, we need to base the evolutionary picture on those pieces that we have access to. Assessing links that can be made between seemingly disparate observations and combining behavioral paradigms with neuroscientific tools in the laboratory will help us continue to unravel the puzzle of what may have happened during language evolution.

5

PRIMATES, CEPHALOPODS, AND THE EVOLUTION OF COMMUNICATION

PETER GODFREY-SMITH

Recent decades have seen dramatic progress in work on animal communication and its evolution, on both empirical and theoretical fronts. Dorothy Cheney and Robert Seyfarth have long been leaders in this research, especially on the empirical side, with their extraordinarily rich studies of communication and social life in vervet monkeys (1990) and baboons (2007). Various theoretical models of communication, developed in different fields, have also begun to cohere in recent years. These models illuminate different facets of the central phenomenon: the coevolution of two kinds of behavior seen in sign use. On one side are behaviors of sign *production*; on the other side are behaviors of sign *interpretation*. Communication comprises the ways these behaviors fit (or fail to fit) together. When a communication system has become established, the sounds, scents, or other marks that an animal makes have been conditioned, through selection, by the patterns of reception and interpretation waiting downstream. The converse is also true: the evolution of patterns of interpretation is an ongoing response to features of sign production. Production and interpretation coevolve.

My term *coevolve* above is understood in a broad way, referring to the shaping of sender and receiver behaviors within a species as well as between them, often within the same agents. Evolution by natural selection, also, is one of a family of processes that can shape and stabilize sign-using behaviors. Other members of this family include reinforcement learning, imitation of successful individuals in a population, and deliberate reflection and choice (Skyrms 2010). These selection processes may operate on their own or in tandem, modulating behavior on different timescales. In the first part of this chapter I'll describe what I take to be an implicit consensus on the theoretical side, though one that leaves many questions unresolved. Recent work in this area has explored the role played by common and conflicting interests, signal cost, iteration of interactions, and the network structures linking communicators (one-on-one interaction versus broadcast to many receivers).[1] Another theme of recent work is the role of combinatorial or syntactic structure in communication systems. Clearly this is an important feature of human language. How widespread is combinatorial structure in animal sign systems, and what sort of transition is involved in achieving it? I'll discuss this topic with particular reference to Seyfarth and Cheney's "The Social Origins of Language." They argue there for significant continuities between human and nonhuman cases, especially in primates. Human and nonhuman primate communication certainly have substantial differences, especially on the production side, but on some central issues, as Seyfarth and Cheney see things, the main transitions come early and the human/nonhuman similarities are deep. This applies to the social function of communication and also to combinatorial structure: "In baboons—and very likely many other primates—vocalizations and social knowledge combine to form a system of communication that is discrete, combinatorial, rule-governed, and open-ended."

After framing the issue of combinatorial structure, I'll argue against some parts of Seyfarth and Cheney's treatment of their own central case, baboons. I'll also make a comparison between baboon communication and a very different signaling system, skin patterning in cephalopods. With respect to some debates about combinatorial structure and complexity in sign use, the two cases are complementary: baboons have simple production and complex interpretation. Cephalopods have complex production and, most likely, simple interpretation. The two cases are flip sides of each other, and in the evolution of combinatorial communication systems, they are both incomplete cases. This comparison casts light on the special features of genuinely combinatorial systems, those in which combinatorial structure is integrated into both the sender and receiver roles.

SENDER-RECEIVER COEVOLUTION

This section sketches a general framework for understanding communication that I take to be supported by a range of models that have been developed, mostly independently, in several different fields.[2] The starting point is the distinction between two roles, which I'll call sometimes *producer* and *interpreter*, and sometimes, more simply, *sender* and *receiver*. These pairs of terms will be used more or less interchangeably. Individuals in an interaction may play one of these roles, or both. The earliest model of the family I have in mind was developed in philosophy, by David Lewis (1969).[3] In the Lewis model, a sender has access to a fact, some information about the world, which might be a feature of the sender itself (such as sex or underlying quality). The sender has access to this fact and sends a message of some kind to a receiver. The receiver acts on the message, in a way that has consequences for both agents. Lewis assumed common interest and common knowledge between sender and receiver, and his model gave a simple account of how rational choice could stabilize the rules of behavior on "each side" of the sign, the rule of production (mapping states of the world to messages) and the rule of interpretation (mapping messages to acts).

Terminologies in this area are diverse. I'm going to use *sign* as a very general term, covering anything that is produced and interpreted in the way covered by the models I'm describing, whether the sign is produced vocally, through gesture, inscription, or in some other way. I'll return to some terminological issues below, but for now please read *sign* very broadly.

Not all communication fits the Lewis pattern, and this is true even before we consider relaxing assumptions of common interest and adding other complexities. An essential feature of the Lewis model is an informational asymmetry between sender and receiver—the "private information," as economists call it, available to the sender—along with an asymmetry involving action. The sender can see the world but not act on it; the receiver can act but can only see the sign. The aim of signaling is then to coordinate the receiver's action with the state of the world: *act-to-state* coordination. Not all communication is like this; sometimes the function of communication is to coordinate one agent's acts with another—*act-to-act* coordination—where the difference between "states" and "acts" is the fact that acts are chosen by one of the agents, while states are determined independently of the strategic choices possible in the game. Much communication in actual settings plays both these roles; actions are coordinated, but in a way conditioned by information about variables whose values are externally determined.

In these mixed cases, in cases where acts are only coordinated with acts, and also in the original cases modeled by Lewis, the heart of the matter is the mutual shaping of senders' and receivers' behaviors, the rules or policies of sign production and sign interpretation.

The simplest models assume common interest between sender and receiver. This is especially clear in the case of the Lewis model, where the sender's messages guide the receiver by reducing uncertainty about the state of the world (carrying information, in Shannon's 1948 sense). It would seem that if the sender and receiver want different acts performed in any given state of the world, then if the sender makes information about this state available to the receiver, the information will be used to produce actions that the sender does not want performed. In such a situation, the sender would have no incentive to signal informatively and hence the receiver no reason to listen. At equilibrium, silence should reign. If this line of argument is accepted, the next question to ask is what happens when there is *partial* common interest between the two agents. That question is the topic of a classic model in economics, due to Crawford and Sobel (1982). They modeled a situation where, roughly speaking, the sender wants to *somewhat* exaggerate their quality (or another relevantly similar state of the world), and the receiver wants not to be taken in by the exaggeration. Overlap of interests was measured by quantifying the *somewhat* in my previous sentence—the sender might want to exaggerate hardly at all (more *common* interest), or a lot (less common interest). If the sender is of quality level X, he or she wants the receiver to act as if the sender were $X+d$, while the receiver prefers to act as if the sender is of quality X; so d then measures the sender's desired exaggeration. Signals that carry some information about the sender's quality can be used in this situation, but Crawford and Sobel showed that as interests diverge, fewer and fewer distinct messages will be used at equilibrium. When interests diverge enough, signaling collapses altogether.

Recent work by Manolo Martínez and me has filled out this picture and added some surprises.[4] These surprises significantly qualify the intuitive verbal argument about the role of common interest given above. We devised a measure of common interest between sender and receiver, called C, that requires weaker assumptions than Crawford and Sobel's and other models. Our measure compares the preference orderings that each agent has over actions that might be produced in each state of the world. There is complete common interest ($C=1$) when sender and receiver agree entirely about their rankings of actions for every state; there is complete conflict of interest when they have reversed orderings in every state ($C=0$). That is, there is complete conflict when in every state of the world, the best action for one agent is the worst for the other. We assumed "cheap talk" (no signal costs) and no iteration of play between agents. Across a large sample of three-state games and using two different methodologies (a static "Nash equilibrium" search and a dynamic model), we found that our measure C is strongly predictive of whether communication can be maintained at all, and of how informative the messages in the system will be. (The "informativeness" of communication is measured as the mutual information between states of the world and the receiver's acts.) We also found surprises; there are cases where informative communication is possible despite complete reversal of preferences in every state of the world ($C=0$).[5] These results show that some commonly made assumptions about the difficulty of maintaining communication in situations of low common interest (with no iteration, no assortment in the population, and no signal costs) are not reliable. However, common interest does make informative communication much easier to maintain.

These results, which use such a simple setting, establish a baseline. Further factors can then help or hinder communication. In biology, since the work of Amotz Zahavi (1975), there has been much exploration of *differential cost* as an enforcer of honesty in signaling. For example, an advertisement of quality can be relied on by a receiver if it is too costly for a low-quality sender to produce. This effect may not be as general an explainer of signal honesty as was once thought (Huttegger, Bruner, and Zollman 2015), but it is one piece of the picture. The role of signal cost had been modeled in economics by Michael Spence (1973), with a very similar message, a few years before Zahavi (1975) sketched his hypothesis.

The sender-receiver models also make more precise a distinction that had been important in the literature for some time, the distinction between *signals* and *cues*. Maynard Smith and Harper, whose 2003 book is an important part of the multidisciplinary literature I'm describing here, define a *signal* as "any act or structure which alters the behavior of other organisms, which evolved because of that effect, and which is effective because the receiver's response has also evolved" (3). A *cue*, in contrast, is something organisms can use to guide their action, but which did not evolve *as* a guide of this kind; it is a byproduct of other processes, or a consequence of fixed physical constraints. Maynard Smith and Harper use the example of a mosquito finding a mammal to bite by tracking CO_2. Carbon dioxide can be used by the mosquito as a cue of the location of a nearby mammal, but it is not a *signal* sent by the mammal. In the terms used here, the production of CO_2 by mammals is not part of a sender's rule that coevolved with the mosquito's use of CO_2 as a "receiver" or "interpreter." The mammal would prefer not to give the mosquito any information about its location, but—as we might say—it can't help doing so. CO_2 is an *unsent* sign. It is produced, but not because of a coevolved sending rule.[6]

The vagueness of my phrase above, "can't help doing so," shows another feature of the situation. If avoiding mosquitoes was sufficiently important to mammals, and some sort of sequestering of CO_2 would keep mosquitoes away, we might imagine a situation in which mammals did evolve such sequestering. In a simple sketch of the mosquito case we assume that producing a trackable plume of CO_2 is a fixed constraint, but it is subject to evolution. There are many cases where the "sending" done by an animal *is* evolving, but in a more constrained and slower way than the "receiving" side is evolving. The other relationship is possible too; Owren, Rendall, and Ryan have recently argued, in effect, that this is seen in some important actual cases of animal communication: a sender can successfully exploit a receiver by making use of biases in the receiver's perceptual and neural mechanisms. The situation is not one in which the receiver *cannot* evolve these mechanisms to counteract the sender's efforts, but, they argue, evolution of these mechanisms on the receiver's side is subject to more constraints.[7]

The cue/signal distinction concerns the role of the sender. In other literatures, *signal* is used to refer to simple signs in which the timing of production is important. The making and use of this book (or its chapters) fits a sender-receiver model, for example, but a book is not usually a signal. In yet another literature, in microbiology, *signal transduction* includes the use of cues as well as signals in the sense above (Lyon 2015). I don't want terminology to be a distraction here, so I'll keep using *sign* in a broad way and sometimes use other terms that should be clear in the immediate context.

ORGANIZED SIGN SYSTEMS AND
COMBINATORIAL STRUCTURE

We might start by asking: what distinguishes the simplest cases, sign systems with no combinatorial structure, from those that do have some? I'll approach this question with a distinction I take to be even more basic, between what I'll call *nominal* and *organized* sign systems.[8] Nominal signs are unstructured in a very strong sense. Not only are they not made up of significant parts—words or similar units—but they are part of a sign system where no natural relations between one sign and another play a communicative role. This term *natural relation* is problematic, but it's the best I have for now.[9] The idea can be illustrated with an example. Consider the classic tale of Paul Revere and the sexton of the Old North Church in Boston in the American Revolution.

The sexton used a lantern code—*one if by land, two if by sea*—to inform Revere of the route of the British attack. This code features a mapping between signs and states of the world, but the difference in magnitude between one and two lanterns does not play any role. One lantern and two lanterns are just distinguishable signals. Compare that case to another. Rather than signaling land versus sea, suppose the sexton used only one lantern, but the brighter the lantern was, the bigger the army he'd seen. Here there is a natural relation between different signs—the *brighter than* relation—that maps to a natural relation between armies—the *larger than* relation. The sexton might instead have used a dimmer lantern for a larger army; that system would work just as well, provided the receiver's rule of interpretation was coordinated with it.

In the case where lantern brightness maps to army size, the sign system is an *organized* one. The actual *one-if-by-land . . .* rule, in contrast, yielded a purely nominal sign system. But in both those cases there is no internal structure in the signs themselves; there is nothing like a syntax. The signs have no internal parts that can be rearranged. An animal alarm call system in which calls are louder (or quieter) when predators are nearer is also a case like this; the sign system is organized even though it has no syntax. Often, though, the *way* an organized sign system is achieved is by means of syntax and internal structure. The signs in the system are related to each other by the *sharing of constituents*, which can be recombined and rearranged. *Bob arrived* and *Bob left* are related by their shared constituent *Bob*. This shared constituent is a feature of the signs that matters to their interpretation. Both say something about a particular individual, Bob.

Sharing a constituent is a natural relation between signs, and it maps to a sharing of constituents between the states of affairs described. Combinatorial structure is one kind of organization in a sign system, one way that signs can be related to each other by communicatively significant transformations. Having parts is a *means* to organization in my sense. There are other means that don't involve internal structure, as in the case where a louder call maps to a closer threat. The important distinction in this area is not whether or not a sign *has* parts. All physical things have parts (at least at this scale). The question is whether the signs' parts have some role in the sender-receiver system, whether the rules of production and interpretation are sensitive to a particular kind of internal structure in the signs.[10]

Suppose the sexton's rule is: show one lantern per British brigade. That is a feature of the sender's rule, and it may or may not be coordinated with the receiver's rule. Revere might have a receiver's rule that takes this into account, or he might not. He might not realize that each lantern says something definite. Similarly, suppose closer predators lead to an animal alarm caller becoming more excited and making a louder call. This is—so far—a sort of inadvertent or de facto organization on the sender's side. It may or may not be picked up in the rule of interpretation used on the receiver's side. We might expect this organization to be quickly made use of by receivers, but it's an open question whether this happens in any particular case. There might be a role for inadvertent or de facto organization on the receiver side, too. A louder call might make the receiver more agitated, just as a result of general features of their perceptual psychology, and this agitation might be a good—or a bad—thing with respect to their response to the call.

Initially these features might be inadvertent, but they may then come to figure in the coevolution of senders' and receivers' behaviors. They might be amplified, suppressed, or transformed. In principle, there can be useable structure in signs that is unused by the receivers. There can also be a situation where structure is *present* in signs, not because of an evolved sender's rule, but by happenstance. This structure, too, might be used or not used by receivers. Suppose the sexton intends to signal in exactly the same way for any sea invasion, but he does not. His alarm call is inadvertently affected by the details, and Revere may or may not pick up on this.

Now I'll combine this with a point made at the end of the previous section. There is another situation where a kind of sign structure arises not by a coevolved sender's rule—not by "design" —but by happenstance. Suppose you hear a lion's roar followed by the bellow of an antelope. You might use this to build a scenario about what's going on.[11] The two pieces, roar and bellow, each play a role. Two roars will be different from one, also (there are two lions to deal with if I go to the waterhole). In cases like this, a structured combination of sounds or other signs makes possible a certain sort of interpretation, but the interpretation is directed at an object whose combinatorial structure is not due to an evolved sender's rule. Instead, there are simpler behaviors of sign production. When they are put together, they yield a structured and interpretable object, but no agent on the sender's side is following a rule of combinatorial sign production. In the lion-antelope case, there is just a useful happenstance combination of simple signs.

These distinctions have gray areas at their boundaries. For example, how do we distinguish a single combinatorially structured sign from a sequence of unstructured, nominal signs from the same sender? Sometimes this is easy, because the parts of the structured sign could not occur on their own. In other cases, the parts might be able to occur on their own, but an argument might be made that their role in a sequence is one that involves genuine combinatorial structure. There's a connection here to a distinction made by Thom Scott-Phillips in two recent articles.[12] He says that there are various cases where two or more animal signals are produced *alongside* one another, and this need not be a "properly combinatorial" system, because in many cases the effect of the sequence is just the sum of the effects of the parts. That is, suppose A_1 is the evolved response to M_1, A_2 is the evolved response to M_2, and if M_1 and M_2 are both sent, the receiver does both A_1 and A_2. This shows, he says, that the system is not a genuine combinatorial one. I agree that there's an important distinction here, without being so sure about some of his judgments about cases. Scott-Phillips says that the honeybee waggle dance is a case where the effect is the sum of the parts, so it's not a genuinely combinatorial system. But if this "sum" talk is to be literally applicable, the parts have to be signs that can be produced, and reacted to, in isolation. In the bee dance, the angle of the dance maps direction and the duration of the runs maps distance. For this to be a sum-of-parts case, it would have to be possible for a bee to dance with a definite direction but no definite duration, and with a definite duration but no direction. At least the latter does make sense, but the former might be doubted. If this separation is not possible, then the angle and duration are more akin to syntactic features of a structured sign. The sentence *Bob arrived* is not the "sum" of *Bob* and *arrived*, in the relevant sense. The word *arrived* cannot achieve anything in isolation, such that we might ask whether this effect is "summed" with the effect of *Bob* when someone interprets *Bob arrived*.

BABOONS AND CEPHALOPODS

With this framework in hand, let's now look at some of the primate behaviors described in Seyfarth and Cheney's "The Social Origins of Language" and elsewhere. The baboons they study live in complex social structures with an important role for ranks. They also make calls. On the production side, there is not a lot of flexibility in what a baboon can do. The repertoire is simple, with about four different calls, and the production rules are stereotypical. But the individuals in a troop can recognize *who* has made a particular call. That means, as Seyfarth and Cheney say, that combinations of signs can carry a lot of information—they carry information in Shannon's sense, as they reduce uncertainty about the state of the world. If you hear a threat-grunt from A followed closely by a scream from B, that is indicative of a particular interaction, one between A and B, and with particular roles. With Seyfarth and Cheney, consider then this sequence: a threat-grunt from a low-ranking individual and a scream of submission from a higher-ranking one. That is a notable combination, a surprising one. It is very different in what it indicates about social affairs than a threat from a high-ranker and submission from a low one. The sequence of a threat from a low-ranking individual and submission from a higher-ranking one indicates that a social reversal or shift has taken place. The combination is informative in that sense, to a hearer. But no agent has the ability to produce a sign with those features, any more than a lion, which can only roar, can tell you what it has attacked. The combination of baboon calls is informative to a sophisticated interpreter, even though there is no coevolved rule of production whose function is producing such signs and making such information available.

I disagree with Seyfarth and Cheney's own description of these cases in their chapter. They say: "In baboons—and very likely many other primates—vocalizations and social knowledge combine to form a system of communication that is discrete, combinatorial, rule-governed, and open-ended." Their basis for saying this in the case of baboons is the sophistication on the receiver side. I think this is not enough, and baboon behavior does not comprise a "system of communication" with combinatorial features, any more than the lion-prey case does.

This case is interesting in the light of the distinction between signals and cues, discussed above. The baboons who call are both signaling; the calls are not mere cues. But the combinatorial structure (such as it is) in what the receiver hears is cue-like. It is a fortuitous consequence of the social ecology and the rules of nominal sign production being followed by individuals. When I say it is "fortuitous," I don't mean it's an accident. The evolution of call production was shaped by the social ecology of baboon life, and this social ecology includes the fact that pairs of calls, as well as individual calls, can be heard. That fact might have been important. But there is no sender anywhere in this system whose behaviors of sign production have been shaped by selection for making calls with combinatorial structure. The structure in the calls is fortuitous in that sense.

There's a contrast between the way Seyfarth and Cheney present their ideas in "The Social Origins of Language" and in their book *Baboon Metaphysics* (2007). In the book, they use data of this kind to make a case for *internal* sophistication in baboons. They argue for a system of internal representation in these animals, for something like a "language of thought" (2007:251). The hypothesis of a *language* of thought might be too strong given the data, as Elisabeth Camp has argued (2009). But the data do support claims of cognitive sophistication and a kind of internal symbolic structure on the interpreter side. In their new work, though, these results are described as showing the presence of a system of communication rather than just a system of internal interpretation. In response, Seyfarth and Cheney might say that the first result does indeed establish the second. Once we've shown that the baboons' way of *assessing* calls has a certain kind of complexity, this shows that the communication *system* itself has that sort of complexity. Their chapter contains passages that suggest this interpretation.[13] I am arguing, however, that with respect to combinatorial structure, it takes two to tango.

Am I merely insisting on one particular way of dividing things up? Suppose they reply: "It's a combinatorial system if the receiver treats it that way." What is wrong with that? I agree there will be many reasonable ways to categorize the cases. But much progress has resulted from focusing on sender-receiver coevolution, and in the light of that framework, a combinatorial system is one with complementary features on each side. There has to be a combinatorial nature to the making of signs, and to their interpretation. The sender constructs a sign with internal structure and the receiver is sensitive to that structure. Cases with complexity on just one side are important in their own right, but they're important as a different sort of phenomenon.

If we look at things this way, we can identify a complementary case, a flip side, to the baboons' combination of features. This is skin patterning in the coleoid cephalopods (octopuses, cuttlefish, and squid). These animals have the ability to change their skin color and pattern in dramatic ways in less than a second. Larger cuttlefish, such as the Australian giant cuttlefish (*Sepia apama*), are probably the most spectacular, especially with respect to colorfulness, but each group has its specialties (Hanlon and Messenger 1996; Darmaillacq, Dickel, and Mather 2014). Octopuses can achieve astonishing camouflage, and squid, as discussed below, are perhaps the most communicative. In all these animals, the color and pattern changes are controlled to a considerable extent by the brain. Their skin contains several color-affecting components. Most importantly, *chromatophores* are sacs of pigment that can be expanded and contracted in precise ways with muscles. Other cells, below the chromatophore layer, reflect ambient light. I'll focus on chromatophores, the most precisely controlled elements in the skin.

The skin of one of these animals contains large numbers of chromatophore units. They can be used to make both static and dynamic patterns, with a huge variety possible. A cuttlefish, for example, has three chromatophore colors, and of the order of a million chromatophores across its skin. Control does not seem to be chromatophore by chromatophore; they tend to work in clumps. But there is still a large number of independently controllable units, and as a result a vast number of patterns possible at a time. Color and pattern can also change rapidly over time.

So, on the production side, there is enormous complexity. What is it for? It is believed that the original function was probably camouflage, and in some species the system has been pressed into a signaling function as well, both intraspecific and interspecific. Some species of cuttlefish have elaborate contests between males, which include displays, and male-female signaling is also common. Octopuses appear to use signaling less than other coleoids (though see Huffard, Caldwell, and Boneka 2010 and Scheel, Godfrey-Smith, and Lawrence 2106). In all these cases, though, it is likely that the interpretation side is vastly simpler than the production side. I'll discuss a possible partial exception below, and in some species there is more complex signaling than in others. But a great deal of combinatorial capacity is probably going unused here, especially on the interpretation side.

The species for which the strongest claims about signal complexity have been made is a reef squid, *Sepioteuthis sepioidea*, in the Caribbean. Martin Moynihan and Arcadio Rodaniche (1982), in a very readable monograph that is an underwater analogue of *Baboon Metaphysics*, argued that these squid employ a "language" on their skin. Reef squid are social, forming shifting groups of six to twelve or more. They have fairly complex courting behaviors, some low-key territoriality, and they also display at predatory fish. Moynihan and Rodaniche charted the combinations of patterns produced and how they were combined with arm positions, and found quite a rich structure. They then argued that squid have a visual language with a syntax. This claim was based mainly on the structure seen in sequences of basic displays, though they also discussed combinations of patterns present at a time.

Among cephalopod biologists these claims of language and syntax have generally been thought too rich. Moynihan and Rodaniche saw too much structure. But what is meant by *too much* here? What determines the "real" amount? In part this is a matter of which patterns are systematically produced, but the other crucial factor is how the patterns are *interpreted* by individuals who see them. Moynihan and Rodaniche were able to chart in some detail the structure of signs produced, but were not able to work out very well their effects on receivers. This is entirely understandable; behavioral observations are difficult with animals of this kind. Squid are skittish and fast-moving, and even a good snorkeler lumbers in comparison.

Moynihan and Rodaniche counted about thirty-one ritualized patterns. They believed that systematic patterns in the sequence with which these displays were produced must have some meaning:

> We cannot, ourselves, in the present state of our knowledge, always and in every case tell the difference in message or meaning between every observed arrangement of particular patterns. We feel, nevertheless, that we must assume that there is a real functional difference of some sort between any two sequences or combinations that can be distinguished from one another. (1982:125)

But the options, by their own lights, seemed limited by the slim range of behaviors seen in intraspecific interactions. They saw occasional territorial defense between groups, many displays directed at predators of other species, and a variety of courtship and sexual behaviors. The variety of displays seems to outrun the variety of responses, and Moynihan and Rodaniche themselves wondered about the possibility of simple explanations for much of what they saw.

The most detailed attempt to follow up Moynihan and Rodaniche's study I know of was done by Jennifer Mather with some collaborators. Mather (2004) discusses a small set of basic communicative displays in *Sepioteuthis*, though some of the displays are graded, the ones discussed aren't claimed to be exhaustive, and the 2004 paper does not consider posture in conjunction with pattern (see also Mather, Griebel, and Byrne 2010).[14] The more complicated exchanges of signals described are preludes to mating. Mather also discusses the difficulty of tracking receivers' interpretations of displays.

It may be that in reef squid there is a hidden role for some of the rich combinatorial structure in displays discussed by Moynihan and Rodaniche. This would probably involve subtle and graded modulation of the basic behaviors associated with aggression and sex. It's possible. We could then ask, as Scott-Phillips does, questions about whether a receiver's responses to combinations of signals have an additive relation to their parts, and so on. In other cephalopods this sort of complex signaling is even more unlikely. Reef squid are more social than other coleiod cephalopods. Octopuses, in particular, are not very social at all, though they produce rich combinations of patterns and color changes, many of which do not, apparently, involve camouflage.

How then should we think about complex pattern production in cephalopods? A partial explanation comes just from noting its origins in camouflage. Camouflage, especially in reef environments, involves producing spatially structured patterns, and that is the likely origin of the pattern-producing machinery. Once pressed into service for communication, in a situation where displays are meant to be seen and understood, rather than *not* seen, the result is a lot of combinatorial capacity on the production side. Some displays made by cephalopods to other species are probably designed to startle the other animal, and these "deimatic" displays are very spatially complex, but intended to have simple results. At least in squid, and perhaps in some other cases, there is probably some genuine combinatorial structure to communicative displays between individuals, but there is probably also a great deal of unused capacity and unattended complexity. The interpretation rules in play are probably not tracking much of the combinatorial detail that is inherent to the production mechanisms. In the squid case, Moynihan and Rodaniche probably did enough to show that the production of combinations is not merely random. This is not so clear in other cases. Offering a speculative hypothesis, I suspect (based on informal observations) that some complex cephalopod displays are nonrandom but also functionless; they are fortuitous reflections of internal processes, byproducts of the close connections between brain and chromatophores, that do not have a comparably complex coevolved interpretation. The complex displays indicate something about the animal, but what is indicated is not being used (much) by normal receivers. Perhaps this is not true, but even if false in all cases, its possibility illustrates how the complexity of sign production can outrun the complexity of interpretation in a system of animal communication.

The cephalopod case is a complement to the baboons, the flip side. In the baboons there is much complexity on the receiver side, but it is aimed at sign structure that is not put in place by any sender. A communication system that is genuinely complex and combinatorial is one in which rich combinatorial structure figures into the rules on *both* sides of the signs, rather than a system in which simple nominal signs are produced but complex interpretations are possible given the social context, and rather than a system with very complex production but where most of the complexity is insignificant to interpreters. Especially in philosophy, but also in scientific discussions, there is a tendency to "choose sides" when giving a theoretical description of communication.

Some people treat communication as a fundamentally expressive phenomenon, and emphasize the sender side (in philosophy, see Grice); other views see communication as a fundamentally interpretive phenomenon, and emphasize the receiver side (in philosophy, see Davidson).[15] The coevolutionary framework shows us that sides should not be chosen.

※   ※   ※

I am grateful to Manolo Martínez and Ron Planer for extensive discussion of these issues.

PART 3

CONCLUSION

Robert M. Seyfarth and
Dorothy L. Cheney

Evolutionary biologists often draw a distinction between the selective forces that cause the early emergence of a trait and those that shape its later elaboration once it has appeared. The evolution of the vertebrate eye from a primitive light receptor provides a classic example (Dawkins 1982). We hope it is clear that our essay focuses exclusively on the earliest stages of language evolution when, we speculate, a communicative system that would eventually become language first began to emerge. We have little to say about the later evolution of language's more complex properties, like case, tense, subject-verb agreement, or the subjunctive, that are certain to have evolved through their own distinct evolutionary process.

The goal of our essay, then, has been to contribute to research on language evolution by asking how evolution has shaped the communication and cognition of animals that live in large social groups. If we can find general rules that specify how social complexity affects communication and communication affects reproductive success—particularly among nonhuman primates—these rules might give us a better idea of the precursors to language and the foundations on which language was built (Cheney and Seyfarth 2007).

THE IMPORTANCE OF PRAGMATICS

Animal communication is embedded in a rich social environment where signalers and recipients interact repeatedly for years at a time, and in which success depends on individuals' ability to recognize each others' relationships and form enduring social bonds (Cheney and Seyfarth 2007; Silk, Alberts, and Altmann 2003; Silk et al. 2009, 2010a, b). In these circumstances, the meaning of a signal depends on both the signal itself and the context in which it is given—and context can include the recognition of another individual's identity, rank, family membership, as well as the recipient's memory of past interactions, not only with the signaler but also with its kin (Cheney and Seyfarth 2007; Marler 1983).

Pragmatics is the subfield of linguistics that examines how context contributes to meaning. By this definition, it seems clear that communication in animals—particularly long-lived, social species like monkeys and apes—constitutes a rich pragmatic system. The ubiquity of pragmatics in animals, combined with the relative scarcity of semantics and syntax, is important for those interested in the evolution of language because it suggests that, as language evolved from prelinguistic systems of communication, semantics and syntax were built upon a foundation of sophisticated pragmatic inference (Seyfarth and Cheney 2016a, b). We have tried, in our original essay, to provide some examples of what pragmatic inference might involve in baboons, with the hope that this work will spur similar studies on other species and prompt ethologists, linguists, neuroscientists, and philosophers to consider what Schlenker et al. (2016) call the "division of labor" between pragmatics, semantics, and syntax, both in animal communication and in the evolution of language. The contributions by Arnold, McWhorter, and Wilson and Petkov in this volume provide good evidence that such research is well underway.

Any system of communication that relies on pragmatics poses a challenging cognitive problem for its participants. To cite a familiar example, when human infants first hear a word that has no meaning for them, they confront a "set of possible meanings that . . . is enormous and technically infinite" (Stevens et al. 2017; see also Fisher and Gleitman 2002; Spelke and Kinzler 2007). How do infants decide which feature of the environment is the best guide to word meaning? To borrow a term from cognitive science, infants confront the *framing* problem. The way they frame an ambiguous utterance—place it in its appropriate context—will determine whether they get the meaning right or wrong (Fisher and Gleitman 2002).

In animals like baboons, the framing problem is especially acute because—unlike human words—the baboons' vocalizations *by themselves* are imprecise in the information they convey. Baboon grunts, for example, are very general, nonspecific signals that occur in a variety of contexts with many different consequences. How do listeners disambiguate these vague signals?

A partial solution arises because, in most cases, communication among animals occurs in stereotypical situations where the variety of possible meanings is sharply reduced (Marler 1983). When a female baboon grunts as she approaches a mother with a new infant, the grunt is unlikely to refer to group movement, the discovery of food, or the approach of a predator; instead, past experience predicts that it almost certainly reflects the approaching female's motivation to touch the infant and interact with the mother.

But even for baboons the framing problem is not always easily solved. Let's assume, for example, that a baboon has recently exchanged aggression with individual A_1 (a member of the A family), and a few minutes later hears a grunt from animal X. In order to respond appropriately, the listener must recognize the identity of animal X, determine whether

X is a member of the A family or some other matriline, and remember whether she has recently interacted in either a friendly or an aggressive manner with any member of A's or X's family. In other words, to determine the grunt's meaning and hence the appropriate response, the listener must draw on all of her knowledge of individual identities and the matrilineal structure of her group, otherwise she's liable to make an inappropriate and maladaptive response. And the listener must do so instantly, many times throughout the day. Doing so, we argue, requires a system of rapid pragmatic inference—one that cannot be based on the slow, deliberate searching among information stored in a large unstructured database. Instead, it requires a set of social concepts (Seyfarth and Cheney 2014b) and hierarchically structured knowledge of the other animals' social relations (Seyfarth and Cheney 2014a, 2016a).

Even among baboons, therefore, the cognitive challenges posed by pragmatic inference are not trivial, and we may assume that challenges increase with social complexity. Nor is there any reason to assume that pragmatic inference is limited to nonhuman primates: we should expect to find it in any species where the information derived from communicative signals depends on a complex interaction between the signal itself and the context in which it occurs (Marler 1983). If we are correct in concluding that pragmatics is widespread in animal communication whereas semantics and syntax are not, it follows that during our evolutionary history the cognitive operations required by pragmatic inference came first, setting the stage for the later evolution of language's more complex formal properties.

SEMANTIC PARITY

Godfrey-Smith (this volume) is right to focus on a fundamental difference between human language and most animal communication. In language, there exists what Fitch (2010) has called "semantic parity" between signaler and recipient, who share a common representational framework for interpreting meaning. Not exactly common, of course, because recipients do not always understand precisely what a signaler means to say—but common enough to get by. By contrast, communicative interactions among animals seem oddly asymmetric: listeners may extract subtle, complex information from a signal but there is little evidence that signalers intend for them to do so (Cheney and Seyfarth 1998). Semantic parity may be largely or entirely absent.

This conclusion, of course, may be premature. Recent experiments on snake alarms by wild chimpanzees (Crockford et al. 2012; Schel et al. 2013) raise the distinct possibility that signalers may modify their calling depending on their perception of whether or not a nearby companion knows about the immediate danger. This, in turn, would require some kind of semantic parity, because it would suggest that signalers both recognize a mismatch between their own and a listener's knowledge and take steps to change this imbalance.

Expanding on the issue of semantic parity, Godfrey-Smith proposes that combinatorial structure provides "one kind of organization in a sign system," one in which signs are "related to each other by communicatively significant transformations." This point is explored in slightly different ways by Schlenker et al. (2016), who note that in many forest monkeys some calls are more informative than others. Diana monkeys (*Cercopithecus diana*), for example, give "alert" calls to a wide variety of stimuli, including mammalian and avian predators, large nonpredatory animals, falling trees, and social disturbances within the group (Gautier and Gautier 1977; Zuberbuhler, Noe, and Seyfarth 1997) but give their "avian alarm" call only to crowned eagles (*Stephanoaetus coronatus*). The avian alarm is, therefore, more precisely informative. Schlenker et al. (2016) posit that "when one call is strictly more informative than another, the most informative one is used whenever possible." If true, this would be interesting in and of itself, but it would also suggest that when a listener assesses the meaning of a particular vocalization, she does so in light of the meaning of other vocalizations in the caller's repertoire. Schlenker et al.'s (2016) detailed analysis of call combinations in forest monkeys provides an interesting complement to Progovac's discussion (this volume) of call combinations.

Godfrey-Smith disagrees with our conclusion that baboon vocal communication constitutes a combinatorial "*system* of communication*" (our italics). He accepts evidence that listeners infer the meaning of a sequence of calls between individuals A and B by combining information about the caller's identities, ranks, and kinship, and by inferring a causal link between A's initial vocalization and B's vocal response. But he notes—as we have ourselves—that this "sequence" arises inadvertently, as an incidental consequence of one animal's signal and another's response. In baboons, the listener is doing all the work. And a true combinatorial system must be one "with complementary features on each side."

Point taken. But one-sided combinatorial thinking shouldn't be dismissed entirely, for at least two reasons. First, as Arnold (this volume) points out, "unintentional behavior plays a systematic role in the types of human signals that are used communicatively and intentionally." Second, for those interested in the evolution of language, the one-sided system of communication among baboons is nonetheless important because, as an evolutionary step along the way to language, it gets us halfway to where we'd like to be. When a baboon listener parses the sequence "A threatens B and B screams" as "A threatens B *and this causes* B to scream," she recognizes an actor (A), an action (threatening), and an acted-upon (B). So while, from the producers' perspective, it's just two individuals interacting, from the listener's perspective it's a sentence. The data support Wilson and Petkov's evidence (this volume) that animals can "process certain relationships between items or elements in sequences of auditory or visual stimuli," and their proposal that these sequences are directly related to nascent grammatical relations.

We conclude, then, that long before our ancestors spoke in sentences, they had a language of thought in which they represented the world—and the meaning of call sequences—in terms of agents, actions, and patients. Long before they could engage in the computations that underlie modern grammar, they performed computations needed to understand the social relations of those around them (for further discussion, see Cheney and Seyfarth 2007, chapter 11).

It's just a guess, but we suspect that the contributions to this volume will not be the last to be written on animal communication and the evolution of language. We hope that our work will draw attention to the rich social context in which nonhuman primate communication occurs; the subtle shades of meaning that context provides; the common ground between pragmatics in language and animal communication; and the ways in which pragmatic inference among animals can be tested through field experiments.

NOTES

WHERE IS CONTINUITY LIKELY TO BE FOUND?, LJILJANA PROGOVAC

1   One reason for Berwick and Chomsky's (2011) proposal that syntax and Merge were initially useful only for thought, but not for communication, has to do with that one person in their evolutionary scenario who got the language mutation. Their argument is that this one person would not have had anybody to communicate with, and that communication could start only much later, after this mutation was passed down through generations. This kind of conundrum arises if you insist, as Berwick and Chomsky do, that language/syntax arose as a single and sudden event/mutation, in its entirety, but not if you envision an incremental, gradualist approach, with precursors (see, e.g., Jackendoff 1999, 2002; Gil 2005, 2012; Progovac 2009a, b, 2015; Hurford 2012; and many others).

2   Even more radically, Berwick and Chomsky's (2011) vague proposal is consistent with claiming that baboons have language comparable to human language, but that their language has not been externalized (yet). For more discussion on this topic, see the review of Berwick and Chomsky (2016) in Progovac (2016b).

3   The idea that a sentence (TP) is built upon the foundation of a small clause is one of the most stable postulates in this syntactic framework, having withstood the test of time and empirical scrutiny. This kind of analysis was originally proposed in Stowell (1981); Burzio (1981); Kitagawa (1985); and further solidified in the work of Koopman and Sportiche (1991); Chomsky (1995); and many others.

4   In addition to these compounds, Progovac (2015, and previous work) identified a host of other "fossil" structures, as well as specific communicative benefits of progressing first from no syntax to the flat two-word stage, and then from this flat stage to the more elaborated hierarchical syntax. In this view, the flat small clause stage provided the necessary stepping stone into hierarchical syntax. In theoretical syntax, the small clause continues to serve as

foundation for building higher layers of structure, as illustrated in (3) in the text.

5 The compounds discussed here are just one representative example and should not be taken as the only example of protosyntax, nor should the reader conclude that insult was the only or primary reason for evolving syntax. Among the verb-noun compounds (6, 8) one also finds those that name plants and animals, which would also have had a survival value. For a detailed discussion of these issues, focusing on both the syntactic and biological side of things, the reader is referred to the relevant chapters in Progovac (2015; see also 2016a for some thoughts on the relevance of cooperation versus competition for the evolution of language).

6 Zuberbuhler (2002) reports that forest monkey communication may include meaningful call combinations even in the wild. Based on the results of experiments conducted in Taï Forest, the author concludes that male Campbell monkeys emit boom calls before their alarm calls in less dangerous situations, such as when there are falling tree branches or distant predators, and that these boom calls may be equivalent to qualifiers such as *maybe*. This would be because the other monkeys do not react to regular alarm calls when they are preceded by such boom calls.

7 According to Tallerman (2012:453–454), "at most we can agree that Kanzi has learned a productive proto-grammar . . . certain properties that we might call proto-syntactic are attested in animal language research. Words can be meaningfully combined, especially in novel ways."

8 As one example, the fitness of lactose tolerance is 2–3 percent higher in dairy areas. It took about 5,000–10,000 years to reach the current rates of lactose tolerance among northern Europeans, which is close to 100 percent with some populations.

9 Yang's focus here was of a different nature, meant to show that there is no ontogeny/phylogeny connection when it comes to language evolution. However, one problem that I see with this study in this respect is that the structures being tested are not comparable, given that articles (*a, the*) are highly abstract functional categories (associated with determiner phrases), late to emerge in children (e.g., Radford 1990), as well as in the grammaticalization processes (e.g., Heine and Kuteva 2007). Articles are not even available in all human languages. That means that children's use of articles already signals a relatively advanced stage of syntactic elaboration, where continuity is not expected to be found.

FLUENCY EFFECTS IN HUMAN
LANGUAGE, JENNIFER ARNOLD

1   While some scholars argue that disfluency is produced intentionally as a signal that the speaker is having trouble (Clark and Fox Tree 2002), there is no theory in which disfluency is a signal of information status per se.

PRIMATES, CEPHALOPODS, AND THE EVOLUTION
OF COMMUNICATION, PETER GODFREY-SMITH

1   For discussions of signal cost, iteration of interactions, and network structures linking communicators—all topics I don't discuss here—see Maynard Smith and Harper (2003), Silk, Kaldor, and Boyd (2000), and Sterelny (2012) respectively.

2   Relevant works in this tradition, besides those discussed in detail in this section, include Millikan (1984), Skyrms (2010), Farrell and Rabin (1996), Zollman, Bergstrom, and Huttegger (2013), Searcy and Nowicki (2005), Godfrey-Smith (2013).

3   Lewis's terminology distinguished *communicator* and *audience*. C. S. Peirce is sometimes seen as the father of this family of ideas, but within my framework here, his is a receiver-focused view. See Godfrey-Smith (2014) for discussion of the Peirce framework and its influence on some recent scientific work.

4   See Godfrey-Smith and Martínez (2013), Martínez and Godfrey-Smith (2016).

5   The C=0 criterion for complete conflict of interest is not as strong as the requirement of a "zero sum" relation between sender and receiver payoffs. See Wagner (2012) for a related model in a zero-sum context.

6   As Maynard Smith and Harper put it, "the crucial point is that the signal must be able to evolve independently of any quality of the signaler [or other variable] about which it conveys information" (2003:4).

7   I discuss Owren, Rendall, and Ryan's (2010) views about exploitation in more detail in Godfrey-Smith (2013). The description of their view given here is what I take to be the most plausible interpretation; sometimes they, like Dawkins and Krebs before them (1978), appear to hold that senders have the upper hand in principle in such interactions. I think there's no reason why this should be the case, and the best way to present their sender-focused view is to do so in the way I have here in the text.

8   This terminology modifies one used by Gallistel and King (2010).

9   Formally, a relation is often identified with a set of ordered pairs (or *n*-tuples). The term *natural* is supposed to strengthen this. This section owes much to discussions with Ron Planer, though he should not be seen as endorsing the analysis.

10 Millikan (1984) may hold that *all* sign systems are organized in my sense, as can be seen by looking closely at (for example) the role of time and place of sign production. I think this is probably not true for all cases, but if this view is right, a category of minimal organization might be distinguished from richer forms.

11 I don't know whether a lion would actually roar when trying to attack an antelope—my African experience is considerably slimmer than Seyfarth and Cheney's. The point could also be made with other examples of interspecific interactions in which calls are made.

12 See Scott-Phillips and Blythe (2013) and Scott-Phillips et al. (2014).

13 "[D]espite their many well-established differences, language and non-human primate communication share a suite of common cognitive operations. Both are discrete, combinatorial systems in which a finite number of signals can generate an infinite number of meanings."

14 I am not sure how to compare the numbers. Mather (2004) discusses four basic communicative body patterns and two concealment ones, but these are not presented as exhaustive. In other work, Mather along with her collaborators distinguishes more displays (Mather, Griebel, and Byrne 2010; Byrne et al. 2003), some of which include body posture as well as skin patterning. I don't know of later studies that recognize the full variety discussed in Moynihan and Rodaniche.

15 Relevant references are Grice (1969) and Davidson (1984).

REFERENCES

INTRODUCTION, MICHAEL L. PLATT

Berwick, R., and N. Chomsky. 2011. The biolinguistic program: The current state of its development. In *The Biolinguistic Enterprise: New Perspectives on the Evolution and Nature of the Human Language Faculty*, edited by A. M. Di Sciullo and C. Boeckx, 19–41. Oxford: Oxford University Press.

Bolhuis, J. J., I. Tattersall, N. Chomsky, and R. C. Berwick. 2014. How could language have evolved? *PLoS Biology* 12:e1001934.

Brannon, E. M., and J. Park. 2015. Phylogeny and ontogeny of mathematical and numerical understanding. In *The Oxford Handbook of Numerical Cognition*, edited by Roi Cohen Kadosh and Ann Dowker, 203–213. Oxford: Oxford University Press.

Brent, L.J.N., S. R. Heilbronner, J. E. Horvath, J. Gonzalez-Martinez, A. V. Ruiz-Lambides, A. Robinson, J.H.P. Skene, and M. L. Platt. 2013. Genetic origins of social networks in rhesus macaques. *Scientific Reports* 3:1042. doi:10.1038/srep01042.

Chang, S. W., J. F. Gariépy, and M. L. Platt. 2013. Neuronal reference frames for social decisions in primate frontal cortex. *Nature Neuroscience* 16:243–250.

Dobzhansky, T. 1973. Nothing in biology makes sense except in the light of evolution. *American Biology Teacher* 35:125–129.

Drayton, L. A., and L. R. Santos. 2016. A decade of theory of mind research on Cayo Santiago: Insights into rhesus macaque social cognition. *American Journal of Primatology* 78:106–116.

Hayden, B. Y., J. M. Pearson, and M. L. Platt. 2009. Fictive reward signals in the anterior cingulate cortex. *Science* 324:948–950.

———. 2011. Neuronal basis of sequential foraging decisions in a patchy environment. *Nature Neuroscience* 14:933–939.

Klein, J., S. V. Shepherd, and M. L. Platt. 2009. Social attention and the brain. *Current Biology* 19:R958–R962. doi:10.1016/j.cub.2009.08.010.

MacLean, E., D. Merritt, and E. M. Brannon. 2008. Social complexity predicts transitive reasoning in prosimian primates. *Animal Behaviour* 76:479–486.

Pearson, J. M., K. K. Watson, and M. L. Platt. 2014. Decision making: The neuroethological turn. *Neuron* 82:950–965.

Platt, M. L., and A. A. Ghazanfar, eds. 2012. *Primate Neuroethology.* Oxford: Oxford University Press.

Santos, L. R., and M. L. Platt. 2014. Evolutionary anthropological insights into neuroeconomics: What non-human primates tell us about human decision-making strategies. In *Neuroeconomics: Decision Making and the Brain,* edited by P. W. Glimcher and E. Fehr, 109–122. London: Academic Press.

Santos, L. R., and A. G. Rosati. 2015. The evolutionary roots of human decision making. *Annual Review of Psychology* 66:321–347.

Sperber, D. 2000. Metarepresentations in an evolutionary perspective. In *Metarepresentations: A Multidisciplinary Perspective,* edited by D. Sperber, 117–137. Oxford: Oxford University Press.

Sperber, D., and D. Wilson. 1986. *Relevance: Communication and Cognition.* Cambridge, MA: Harvard University Press.

Templer, V. L., and R. R. Hampton. 2013. Episodic memory in nonhuman animals. *Current Biology* 23:R801–R806.

Toda, K., and M. L. Platt. 2015. Animal cognition: Monkeys pass the mirror test. *Current Biology* 25:R64–R66.

THE SOCIAL ORIGINS OF LANGUAGE, ROBERT M. SEYFARTH AND DOROTHY L. CHENEY

Adachi, I., and R. R. Hampton. 2012. Rhesus monkeys see who they hear: Spontaneous cross-modal memory for familiar conspecifics. *PLoS One* 6:e23345.

Barsalou, L., W. K. Simmons, A. K. Barbey, and C. D. Wilson. 2003. Grounding conceptual knowledge in modality-specific systems. *Trends in Cognitive Sciences* 7:84–91.

Beecher, M. D. 1989. Signaling systems for individual recognition: An information theory approach. *Animal Behaviour* 38:248–261.

Belin, P., and D. Zattore. 2003. Adaptation to speaker's voice in right anterior temporal lobe. *Neuroreport* 14:2105–2109.

Belin, P., R. Zattore, P. Lafaille, P. Ahad, and B. Pike. 2000. Voice-selective areas in human auditory cortex. *Nature* 403:309–312.

Bergman, T., J. C. Beehner, D. L. Cheney, and R. M. Seyfarth. 2003. Hierarchical classification by rank and kinship in baboons. *Science* 302:1234–1236.

Bickerton, D. 1990. *Language and Species.* Chicago: University of Chicago Press.

Cantlon, J. F., and E. Brannon. 2007. Basic math in monkeys and college students. *PLoS Biology* 5:e328.

Cheney, D. L., R. A. Palombit, and R. M. Seyfarth. 1996. The function and mechanisms underlying baboon contact barks. *Animal Behaviour* 52: 507–518.

Cheney, D. L., and R. M. Seyfarth. 1990. *How Monkeys See the World: Inside the Mind of Another Species*. Chicago: University of Chicago Press.

———. 1998. Why monkeys don't have language. In *The Tanner Lectures on Human Values*, vol. 19, edited by G. Petersen, 175–219. Salt Lake City: University of Utah Press.

———. 1999. Recognition of other individuals' social relationships by female baboons. *Animal Behaviour* 58:67–75.

———. 2007. *Baboon Metaphysics: The Evolution of a Social Mind*. Chicago: University of Chicago Press.

Cheney, D. L., R. M. Seyfarth, and J. B. Silk. 1995a. The responses of female baboons (*Papio cynocephalus ursinus*) to anomalous social interactions: Evidence for causal reasoning? *Journal of Comparative Psychology* 109:134–141.

———. 1995b. The role of grunts in reconciling opponents and facilitating interactions among female baboons. *Animal Behaviour* 50:249–257.

Clark, H. 1996. *Using Language*. Cambridge: Cambridge University Press.

Clutton-Brock, T. H., and S. Albon. 1979. The roaring of red deer and the evolution of honest advertisement. *Behaviour* 69:145–170.

Clutton-Brock, T. H., S. D. Albon, R. M. Gibson, and F. E. Guinness. 1979. The logical stag: Adaptive aspects of fighting in red deer. *Animal Behaviour* 27:211–225.

Crockford, C., R. M. Wittig, R. Mundry, and K. Zuberbuhler. 2012. Wild chimpanzees inform ignorant group members of danger. *Current Biology* 22:142–146.

Davies, N. B., and T. R. Halliday. 1978. Deep croaks and fighting assessment in toads, Bufo bufo. *Nature* 275:683–685.

de Waal, F.B.M., and L. Ferrari. 2010. Towards a bottom-up perspective on animal and human cognition. *Trends in Cognitive Sciences* 14:201–207.

Enard, W., M. Przeworski, S. E. Fisher, C.S.L. Lai, V. Wiebe, T. Kitano, A. P. Monaco, and S. Paabo. 2002. Molecular evolution of FOXP2: A gene involved in speech and language. *Nature* 418:869–872.

Engh, A.L., R. R. Hoffmeier, D. L. Cheney, and R. M. Seyfarth. 2006. Who, me? Can baboons infer the target of a vocalization? *Animal Behaviour* 71:381–387.

Feigenson, L., S. Dehaene, and E. Spelke. 2004. Core systems of number. *Trends in Cognitive Sciences* 8:307–314.

Ferrari, L., L. Bonini, and L. Fogassi. 2009. From monkey mirror neurons to mirror related behaviours: Possible direct and indirect pathways. *Philosophical Transactions of the Royal Society B* 365:2311–2323.

Fischer, J., M. Metz, D. L. Cheney, and R. M. Seyfarth. 2001. Baboon responses to graded bark variants. *Animal Behaviour* 61:925–931.

Fitch, W. T. 2010. *The Evolution of Language*. Cambridge: Cambridge University Press.

Freiwald, W., D. Y. Tsao, and M. S. Livingston. 2009. A face feature space in the macaque temporal lobe. *Nature Neuroscience* 12:1187–1196.

Ghazanfar, A. A., C. Chandrasekaran, and N. K. Logothetis. 2008. Interaction between the superior temporal sulcus and auditory cortex mediate dynamic face/voice integration in rhesus monkeys. *Journal of Neuroscience* 28:4457–4469.

Ghazanfar, A. A., and N. Logothetis. 2003. Facial expressions linked to monkey calls. *Nature* 423:937–938.

Ghazanfar, A. A., J. X. Maier, K. L. Hoffman, and N. Logothetis. 2005. Multisensory integration of dynamic faces and voices in rhesus monkey auditory cortex. *Journal of Neuroscience* 25:5004–5012.

Gifford, G. W., K. A. MacLean, M. D. Hauser, and Y. A. Cohen. 2005. The neurophysiology of functionally meaningful categories: Macaque ventrolateral prefrontal cortex plays a critical role in spontaneous categorization of species-specific vocalizations. *Journal of Cognitive Neuroscience* 17:1471–1482.

Gil da Costa, R., A. Braun, M. Lopes, M. D. Hauser, R. E. Carson, P. Herskovitch, and A. Martin. 2004. Towards an evolutionary perspective on conceptual representation: Species-specific calls activate visual and affective processing systems in the macaques. *Proceedings of the National Academy of Sciences USA* 101:17516–17521.

Gilbey, I. C., L.J.N. Brent, E. E. Wroblewski, R. Rudicell, B. Hahn, J. Goodall, and A. E. Pusey. 2012. Firness benefits of coalitionary aggression in male chimpanzees. *Behavioral Ecology and Sociobiology* 67:373–381.

Grafen, A. 1990a. Sexual selection unhandicapped by the Fischer process. *Journal of Theoretical Biology* 144:473–516.

———. 1990b. Biological signals as handicaps. *Journal of Theoretical Biology* 144:517–546.

Hammerschmidt, K., and J. Fischer. 2008. Constraints in primate vocal production. In *The Evolution of Communicative Creativity: From Fixed Signals to Contextual Flexibility*, edited by U. Griebel and K. Oller, 93–120. Cambridge, MA: MIT Press.

Hauser, M. D., N. Chomsky, and W. Fitch. 2002. The faculty of language: What is it, who has it, and how did it evolve? *Science* 298:1569–1579.

Hurford, J. 1990. Beyond the roadblock in linguistic evolution studies. *Behavioral and Brain Sciences* 13:736–737.

———. 1998. Introduction: The emergence of syntax. In *Approaches to the Evolution of Language*, edited by J. R. Hurford, M. Studdert-Kennedy, and C. Knight, 299–304. Cambridge: Cambridge University Press.

———. 2003. The neural basis of predicate-argument structure. *Behavioral and Brain Sciences* 26:261–316.

———. 2007. *The Origins of Meaning*. Oxford: Oxford University Press.

Jackendoff, R. 1987. *Consciousness and the Computational Mind*. New York: Basic Books.

———. 1994. *Patterns in the Mind*. New York: Basic Books.

———. 2002. *Foundations of Language: Brain, Meaning, Grammar, Evolution*. Oxford: Oxford University Press.

Jacobs, L. F., and F. Schenk. 2003. Unpacking the cognitive map: The parallel map theory of hippocampal function. *Psychological Review* 110:285–315.

Jordan, K. E., E. L. MacLean, and E. Brannon. 2008. Monkeys match and tally quantities across senses. *Cognition* 108:617–625.

Kanwisher, N., J. McDermott, and M. M. Chun. 1997. The fusiform face area: A module in human extrastriate cortex specialized for face perception. *Journal of Neuroscience* 17:4302–4311.

Kirby, S. 1998. Fitness and the selective advantage of language. In *Approaches to the Evolution of Language*, edited by J. R. Hurford, M. Studdert-Kennedy, and C. Knight, 359–383. Cambridge: Cambridge University Press.

Kitchen, D. M., D. L. Cheney, and R. M. Seyfarth. 2003. Male chacma baboons (*Papio hamadryas ursinus*) discriminate loud call contests between rivals of different relative ranks. *Animal Cognition* 8:1–6.

Kuhl, P., and A. N. Meltzoff. 1984. The intermodal representation of speech in infants. *Infant Behavior and Development* 7:361–381.

Laidre, M., and R. Johnstone. 2013. Animal signals. *Current Biology* 23:R829–R833.

Martin, A. 1998. Organization of semantic knowledge and the origin of words in the brain. In *The Origin and Diversification of Language*, edited by N. G. Jablonski and L. C. Aiello, 69–88. San Francisco: California Academy of Sciences.

Menzel, R. 2011. Navigation and communication in honeybees. In *Animal Thinking: Contemporary Issues in Comparative Cognition*, edited by R. Menzel and J. Fischer, 9–22. Cambridge, MA: MIT Press.

Newmeyer, F. 1991. Functional explanations in linguistics and the origins of language. *Language and Communication* 11:3–28.

———. 2003. Grammar is grammar and usage is usage. *Language* 79: 682–707.

Ouattara, K., A. Lemasson, and K. Zuberbuhler. 2009. Campbell's monkeys concatenate vocalizations into context-specific call sequences. *Proceedings of the National Academy of Sciences USA* 106:22026–22031.

Owren, M. J., R. M. Seyfarth, and D. L. Cheney. 1997. The acoustic features of vowel-like grunt calls in chacma baboons (*Papio cynocephalus ursinus*): Implications for production processes and functions. *Journal of the Acoustic Society of America* 101:2951–2963.

Patterson, M. L., and J. F. Werker. 2003. Two-month old infants match phonetic information in lips and voice. *Developmental Science* 6:191–196.

Petkov, C. I., C. Kayser, T. Steudel, K. Whittingstall, M. Augath, and N. K. Logothetis. 2008. A voice region in the monkey brain. *Nature Neuroscience* 11:367–374.

Pinker, S. 1994. *The Language Instinct*. New York: W. W. Morrow.

Pinker, S., and P. Bloom. 1990. Natural language and natural selection. *Behavioral and Brain Science* 13:713–738.

Pravosudov, V., and N. S. Clayton. 2002. A test of the adaptive specialization hypothesis: Population differences in caching, memory, and the hippocampus in black-capped chickadees (*Poecile atricapilla*). *Behavioral Neuroscience* 116:515–522.

Premack, D. 1976. *Intelligence in Ape and Man*. Hillsdale, NJ: Lawrence Erlbaum Associates.

Reby, D., K. McComb, B. Cargnelutti, C. Darwin, W. T. Fitch, and T. H. Clutton-Brock. 2005. Red deer stags use formants as assessment cues during intrasexual agonistic interactions. *Proceedings of the Royal Society B* 272:941–947.

Rendall, D., D. L. Cheney, and R. M. Seyfarth. 2000. Proximate factors mediating "contact" calls in adult females and their infants. *Journal of Comparative Psychology* 114:36–46.

Roth, T. C., L. D. LaDage, C. A. Freas, and V. Pravosudov. 2011. Variation in memory and the hippocampus across populations from different climates: A common garden approach. *Proceedings of the Royal Society B* 279:402–410.

Roth, T. C., and V. Pravosudov. 2009. Hippocampal volume and neuron numbers increase along a gradient of environmental harshness: A large-scale comparison. *Proceedings of the Royal Society B* 276:401–405.

Savage-Rumbaugh, E. S., D. M. Rumbaugh, S. T. Smith, and J. Lawson. 1980. Reference: The linguistic essential. *Science* 210:922–925.

Schel, A. M., Z. Machanda, S. W. Townsend, K. Zuberbuhler, and K. E. Slocombe 2013. Chimpanzee food calls are directed at specific individuals. *Animal Behaviour* 86:955–965.

Schel, A. M., S. W. Townsend, Z. Machanda, K. Zuberbuhler, and K. E. Slocombe. 2013. Chimpanzee alarm call production meets key criteria for intentionality. *PLoS One* 8:e76674.

Schino, G., R. Ventura, and A. Troisi. 2005. Grooming and aggression in captive Japanese macaques. *Primates* 46:207–209.

Searcy, W. A., and S. Nowicki. 2005. *The Evolution of Communication.* Princeton: Princeton University Press.

Seyfarth, R. M., and D. L. Cheney. 2003. Signalers and receivers in animal communication. *Annual Review of Psychology* 54:145–173.

———. 2010. Production, usage, and comprehension in animal vocalizations. *Brain and Language* 115:92–100.

———. 2013. The evolution of concepts about agents. In *Navigating the Social World: What Infants, Children, and Other Species Can Tell Us,* edited by M. R. Banaji and S. A. Gelman, 27–30. Oxford: Oxford University Press.

———. 2014. The evolution of concepts about agents: Or, what do animals recognize when they recognize an individual? In *Concepts: New Directions,* edited by E. Margolis, 57–76. Cambridge, MA: MIT Press.

Seyfarth, R. M., D. L. Cheney, T. Bergman, J. Fischer, K. Zuberbuhler, and K. Hammerschmidt. 2010. The central importance of information in studies of animal communication. *Animal Behaviour* 80:3–8.

Silk, J. B. 1999. Male bonnet macaques use information about third-party rank relationships to recruit allies. *Animal Behaviour* 58:45–51.

Silk, J. B., S. C. Alberts, and J. Altmann. 2003. Social bonds of female baboons enhance infant survival. *Science* 302:1231–1234.

Silk, J. B., J. Altmann, and S. C. Alberts. 2006a. Social relationships among adult female baboons (*Papio cynocephalus*): I. Variation in the strength of social bonds. *Behavioral Ecology and Sociobiology* 61:183–195.

———. 2006b. Social relationships among adult female baboons (*Papio cynocephalus*): II. Variation in the quality and stability of social bonds. *Behavioral Ecology and Sociobiology* 61:197–204.

Silk, J. B., J. C. Beehner, T. Bergman, C. Crockford, A. L. Engh, L. Moscovice, R. M. Wittig, R. M. Seyfarth, and D. L. Cheney. 2009. The benefits of social capital: Close social bonds among female baboons enhance offspring survival. *Proceedings of the Royal Society B* 276: 3099–3104.

———. 2010. Strong and consistent social bonds enhance the longevity of female baboons. *Current Biology* 20:1359–1361.

Silk, J. B., E. Kaldor, and R. Boyd. 2000. Cheap talk when interests conflict. *Animal Behaviour* 59:423–432.

Silk, J. B., D. Rendall, D. L. Cheney, and R. M. Seyfarth. 2003. Natal attraction in adult female baboons (*Papio cynocephalus ursinus*) in the Moremi Game Reserve, Botswana. *Ethology* 109:627–644.

Skyrms, B. 2010. *Signals: Evolution, Learning, and Information.* Oxford: Oxford University Press.

Sliwa, J., J-R. Duhamel, O. Pascalis, and S. Wirth. 2011. Spontaneous voice-face identity matching by rhesus monkeys for familiar conspecifics and humans. *Proceedings of the National Academy of Sciences USA* 108:1735–1740.

Steiper, M. E., N. M. Young, and T. Y. Sukarna. 2004. Genomic data support the hominoid slowdown and an early Oligocene estimate for the hominoid-cercopithecoid divergence. *Proceedings of the National Academy of Sciences USA* 101:17021–17026.

Tsao, G. Y., W. A. Freiwald, T. A. Knutsen, J. B. Mandeville, and R. B. Tootell. 2003. Faces and objects in macaque cerebral cortex. *Nature Neuroscience* 6:989–995.

Tsao, G. Y., W. A. Freiwald, R. B. Tootell, and M. S. Livingston. 2006. A cortical region consisting entirely of face-sensitive cells. *Science* 311:670–674.

Van Lancker, D. R., J. L. Cummings, J. Kreiman, and B. H. Dobkin. 1988. Phonagnosia: A dissociation between familiar and unfamiliar voices. *Cortex* 24:195–209.

Wilkinson, G., and G. N. Dodson. 1997. Function and evolution of antlers and eye stalks in flies. In *The Evolution of Mating Systems in Insects and Arachnids*, edited by J. C. Choe and B. J. Crespi, 310–328. Cambridge: Cambridge University Press.

Wittig, R. M., C. Crockford, E. Ekberg, R. M. Seyfarth, and D. L. Cheney. 2007. Kin-mediated reconciliation substitutes for direct reconciliation in female baboons. *Proceedings of the Royal Society B* 274:1109–1115.

Wittig, R.M., C. Crockford, R.M. Seyfarth, and D. L. Cheney. 2007. Vocal alliances in chacma baboons. *Behavioral Ecology and Sociobiology* 61:899–909.

Worden, R. 1998. The evolution of language from social intelligence. In *Approaches to the Evolution of Language*, edited by J. R. Hurford, M. Studdert-Kennedy, and C. Knight, 148–168. Cambridge: Cambridge University Press.

Wright, T. M., K. A. Pelphrey, T. Allison, M. J. McKeown, and G. McCarthy. 2003. Polysensory interactions along lateral temporal regions evoked by audio-visual speech. *Cerebral Cortex* 13:1034–1043.

Yee, E., D. Drucker, and S. L. Thompson-Schill. 2010. fMRI-adaptation evidence of overlapping neural representations for objects related in function or manipulation. *Neuroimage* 50:753–763.

Zuberbuhler, K. 2014. Experimental field studies with non-human primates. *Current Opinion in Neurobiology* 28:150–156.

Zuberbuhler, K., D. L. Cheney, and R. M. Seyfarth. 1999. Conceptual semantics in a nonhuman primate. *Journal of Comparative Psychology* 113:33–42.

LINGUISTICS AND PRAGMATICS, JOHN MCWHORTER

Bickerton, D. 1990. *Language and Species*. Chicago: University of Chicago Press.

———. 2009. *Adam's Tongue: How Humans Made Language, How Language Made Humans*. New York: Hill and Wang.

Brinton, L. 1996. *Grammaticalization and Discourse Functions*. Berlin: Mouton de Gruyter.

Culicover, P., and R. Jackendoff. 2005. *Simpler Syntax*. New York: Oxford University Press.

D'Arcy, A. 2007. *Like* and language ideology: Disentangling fact from fiction. *American Speech* 81:386–419.

Dunbar, R. 1998. *Grooming, Gossip and the Origin of Language*. Cambridge, MA: Harvard University Press.

Falk, D. 2009. *Finding Our Tongues*. New York: Basic Books.

Hauser, M., N. Chomsky, and T. Fitch. 2002. The faculty of language: What is it, who has it, and how did it evolve? *Science* 298:1569–1579.

Jackendoff, R. 2002. *Foundations of Language: Brain, Meaning, Grammar, Evolution*. New York: Oxford University Press.

Lupyan, G., and R. Dale. 2010. Language structure is partly determined by social structure. *PLoS One* 5:1.

Matthews, S., and V. Yip. 1994. *Cantonese: A Comprehensive Grammar*. London: Routledge.

McWhorter, J. H. 2009. Oh, Nɔɔ! A bewilderingly multifunctional Saramaccan word teaches us how a creole language develops complexity. In *Language Complexity as an Evolving Variable*, edited by G. Sampson, D. Gil, and P. Trudgill, 141–163. Oxford: Oxford University Press.

Mithen, S. J. 2005. *The Singing Neanderthals*. Cambridge, MA: Harvard University Press.

Scott-Phillips, T. 2015. *Speaking Our Minds*. New York: Palgrave Macmillan.

Traugott, E. C., and R. Dasher. 2005. *Regularity in Semantic Change*. Cambridge: Cambridge University Press.

WHERE IS CONTINUITY LIKELY TO BE FOUND?, LJILJANA PROGOVAC

Berwick, R., and N. Chomsky. 2011. The biolinguistic program: The current state of its development. In *The Biolinguistic Enterprise: New Perspectives on the Evolution and Nature of the Human Language Faculty*, edited by A. M. Di Sciullo and C. Boeckx, 19–41. Oxford: Oxford University Press.

———. 2016. *Why Only Us? Language and Evolution*. Cambridge, MA: MIT Press.

Bolhuis, J. J., I. Tattersall, N. Chomsky, and R. C. Berwick. 2014. How could language have evolved. *PLoS Biology*. doi:10.1371/journal.pbio.1001934.

Burling, R. 2005. *The Talking Ape: How Language Evolved*. Oxford: Oxford University Press.

Burzio, L. 1981. Intransitive verbs and Italian auxiliaries. Ph.D. dissertation, Massachusetts Institute of Technology.

Cheney, D. L., and R. M. Seyfarth. 2007. *Baboon Metaphysics: The Evolution of a Social Mind*. Chicago: Chicago University Press.

Chomsky, N. 1995. *The Minimalist Program*. Cambridge, MA: MIT Press.

———. 2010. Some simple evo-devo theses: How true might they be for language? In *Approaches to the Evolution of Language*, edited by R. K. Larson, V. M. Deprez, and H. Yamakido, 45–62. Cambridge: Cambridge University Press.

Code, C. 1982. Neurolinguistic analysis of recurrent utterances in aphasia. *Cortex* 18:141–152.

———. 2005. First in, last out? The evolution of aphasic lexical speech automatisms to agrammatism and the evolution of human communication. *Interaction Studies* 6:311–334.

Darwin, C. 1872. *The Expression of the Emotions in Man and Animals*. London: John Murray.

———. 1874. *The Descent of Man, and Selection in Relation to Sex*. New edition, revised and augmented. New York: Hurst and Company.

Fitch, W. T. 2010. *The Evolution of Language*. Cambridge: Cambridge University Press.

Gardner, R. A., B. T. Gardner, and T. E. Van Cantfort. 1989. *Teaching Sign Language to Chimpanzees*. Albany: SUNY Press.

Gil, D. 2005. Isolating-monocategorial-associational language. In *Hand-*

book of Categorization in Cognitive Science, edited by H. Cohen and C. Lefebvre, 347–379. Amsterdam: Elsevier.

———. 2012. Where does predication come from? Canadian Journal of Linguistics 57:303–333.

Greenfield, P. M., and S. Savage-Rumbaugh. 1990. Language and intelligence in monkeys and apes. In Grammatical Combination in Pan paniscus: Process of Learning and Invention in the Evolution and Development of Language, edited by S. T. Parker and K. R. Gibson, 540–579. Cambridge: Cambridge University Press.

Heine, B., and T. Kuteva. 2007. The Genesis of Grammar: A Reconstruction. Oxford: Oxford University Press.

Hurford, J. R. 2012. The Origins of Grammar: Language in the Light of Evolution II. Oxford: Oxford University Press.

Jackendoff, R. 1999. Possible stages in the evolution of the language Capacity. Trends in Cognitive Sciences 3:272–279.

———. 2002. Foundations of Language: Brain, Meaning, Grammar, Evolution. Oxford: Oxford University Press.

Kitagawa, Y. 1985. Small but clausal. Chicago Linguistic Society 21: 210–220.

Koopman, H., and D. Sportiche. 1991. The position of subjects. Lingua 85:211–258.

Miller, G. A. 2000. The Mating Mind: How Sexual Choice Shaped the Evolution of Human Nature. London: William Heinemann.

Patterson, F., and W. Gordon. 1993. The case for the personhood of gorillas. In The Great Ape Project, edited by P. Cavalieri and P. Singer, 58–77. New York: St. Martin's Griffin.

Progovac, L. 2009a. Layering of grammar: Vestiges of proto-syntax in present-day languages. In Language Complexity as an Evolving Variable, edited by G. Sampson, D. Gil, and P. Trudgill, 203–212. Oxford: Oxford University Press.

———. 2009b. Sex and syntax: Subjacency revisited. Biolinguistics 3.2–3:305–336.

———. 2014. Degrees of complexity in syntax: A view from evolution. In Measuring Grammatical Complexity, edited by Frederick J. Newmeyer and Laurel B. Preston, 83–102. Oxford: Oxford University Press.

———. 2015. Evolutionary Syntax. Oxford: Oxford University Press. Oxford Scholarship Online.

———. 2016a. A gradualist scenario for language evolution: Precise linguistic reconstruction of early human and (Neanderthal) grammars. Frontiers in Psychology 7:1714. doi:10.3389/fpsyg.2016.01714.

———. 2016b. Review of Why Only Us? Language and Evolution, by R. C. Berwick and N. Chomsky. Language 92.4:992–996.

Progovac, L., and J. L. Locke. 2009. The urge to merge: Ritual insult and the evolution of syntax. *Biolinguistics* 3.2–3:337–354.

Radford, A. 1990. *Syntactic Theory and the Acquisition of English Syntax*. Oxford: Blackwell.

Savage-Rumbaugh, E. S., and R. Lewin. 1994. *Kanzi: The Ape at the Brink of the Human Mind*. New York: John Wiley and Sons.

Stone, L., and P. F. Lurquin. 2007. *Genes, Culture, and Human Evolution: A Synthesis*. Blackwell.

Stowell, T. 1981. Origins of phrase structure. Ph.D. dissertation, Massachusetts Institute of Technology.

Tallerman, M. 2012. What is syntax? In *The Oxford Handbook of Language Evolution*, edited by M. Tallerman and K. R. Gibson, 442–455. Oxford: Oxford University Press.

Yang, C. 2013. Ontogeny and phylogeny of language. *Proceedings of the National Academy of Sciences of the USA*. doi:10.1073/pnas.121680 3110 PNAS.

Zuberbuhler, K. 2002. A syntactic rule in forest monkey communication. *Animal Behaviour* 63:293–299.

FLUENCY EFFECTS IN HUMAN LANGUAGE, JENNIFER E. ARNOLD

Arnold, J. E. 1998. Reference form and discourse patterns. Ph.D. dissertation, Stanford University.

———. 2008. THE BACON not the bacon: How children and adults understand accented and unaccented noun phrases. *Cognition* 108:69–99. doi:10.1016/j.cognition.2008.01.001.

———. 2016. Explicit and emergent mechanisms of information status. *Topics in Cognitive Science* 8:722–736. doi:10.1111/tops.12220.

Arnold, J. E., J. G. Eisenband, S. Brown-Schmidt, and J. C. Trueswell. 2000. The rapid use of gender information: Evidence of the time course for pronoun resolution from eyetracking. *Cognition* 76:B13–B26. doi:10.1016/S0010-0277(00) 00073-1.

Arnold, J. E., C. Hudson Kam, and M. K. Tanenhaus. 2007. If you say thee uh you are describing something hard: The on-line attribution of disfluency during reference comprehension. *Journal of Experimental Psychology: Learning, Memory, and Cognition* 33:914–930. doi:10 .1037/0278-7393.33.5.914.

Arnold, J. E., J. M. Kahn, and G. Pancani. 2012. Audience design affects acoustic reduction via production facilitation. *Psychonomic Bulletin and Review* 19:505–512. doi:10.3758/s13423-012-0233-y.

Arnold, J. E., and S. C. Lao. 2008. Put in last position something previously unmentioned: Word order effects on referential expectancy and reference comprehension. *Language and Cognitive Processes* 23:282–295. doi:10.1080/01690960701536805.

Arnold, J. E., G. C. Pancani, and E. C. Rosa. 2015. Acoustic prominence is perceived differently for fluent and distracted speakers. Unpublished manuscript, University of North Carolina at Chapel Hill.

Arnold, J. E., and M. K. Tanenhaus. 2011. Disfluency isn't just um and uh: The role of prosody in the comprehension of disfluency. In *The Processing and Acquisition of Reference*, edited by E. Gibson and N. Perlmutter, 197–217. Cambridge, MA: MIT Press.

Arnold, J. E., M. K. Tanenhaus, R. Altmann, and M. Fagnano. 2004. The old and thee, uh, new. *Psychological Science* 15:578–582.

Arnold, J. E., T. Wasow, A. Asudeh, and P. Alrenga. 2004. Avoiding attachment ambiguities: The role of constituent ordering. *Journal of Memory and Language* 51:55–70. doi:10.1016/j.jml.2004.03.006.

Arnold, J. E., and D. G. Watson. 2015. Synthesizing meaning and processing approaches to prosody: Performance matters. *Language, Cognition, and Neuroscience* 30:88–102. doi:10.1080/01690965.2013.840733.

Bard, E. G., A. H. Anderson, C. Sotillo, M. Aylett, G. Doherty-Sneddon, and A. Newlands. 2000. Controlling the intelligibility of referring expressions in dialogue. *Journal of Memory and Language* 42:1–22. doi:10.1006/jmla.1999.2667.

Bard, E. G., and M. P. Aylett. 2004. Referential form, word duration, and modeling the listener in spoken dialogue. In *Approaches to Studying World-Situated Language Use: Bridging the Language-as-product and Language-as-action Traditions*, edited by J. C. Trueswell and M. K. Tanenhaus, 173–191. Cambridge, MA: MIT Press.

Bell, A., J. M. Brenier, M. Gregory, C. Girand, and D. Jurafsky. 2009. Predictability effects on durations of content and function words in conversational English. *Journal of Memory and Language* 60:92–111. doi:10.1016/j.jml.2008.06.003.

Bell, A., D. Jurafsky, E. Fosler-Lussier, C. Girand, M. Gregory, and D. Gildea. 2003. Effects of disfluencies, predictability, and utterance position on word form variation in English conversation. *Journal of the Acoustical Society of America* 113:1001–1024. doi:10.1121/1.1534836.

Bock, J. K., and D. E. Irwin. 1980. Syntactic effects of information availability in sentence production. *Journal of Verbal Learning and Verbal Behavior* 19:467–484.

Bradlow, A. R., G. Torretta, and D. B. Pisoni. 1996. Intelligibility of normal speech I: Global and fine-grained acoustic-phonetic talker characteristics. *Speech Communication* 20:255–272.

Breen, M., E. Fedorenko, M. Wagner, and E. Gibson. 2010. Acoustic correlates of information structure. *Language and Cognitive Processes* 25:1044–1098. doi:10.1080/01690965. 2010.504378.

Brown, G. 1983. Prosodic structure and the given/new distinction. In *Prosody: Models and Measurements*, edited by A. Cutler and D. R. Ladd, 67–77. New York: Springer-Verlag.

Bybee, J., and P. Hopper. 2001. Introduction to frequency and emergence of linguistic structure. In *Frequency and the Emergence of Linguistic Structure*, edited by J. Bybee and P. Hopper, 1–24. Amsterdam: John Benjamins.

Chafe, W. 1976. Givenness, contrastiveness, definiteness, subjects, topics, and point of view. In *Subject and Topic*, edited by C. N. Li, 25–56. New York: Academic Press.

Cheney, D. L., and R. M. Seyfarth. 1997. Animals don't have language. Tanner Lectures on Human Values, delivered at Cambridge University, March 10–12.

Christodoulou, A. 2012. Variation in word duration and planning. Ph.D. dissertation, University of North Carolina at Chapel Hill.

Clark, H. H. 1996. *Using Language*. Cambridge: Cambridge University Press.

Clark, H. H., and J. E. Fox Tree. 2002. Using uh and um in spontaneous speaking. *Cognition* 84:73–111.

Dahan, D., M. K. Tanenhaus, and C. G. Chambers. 2002. Accent and reference resolution in spoken language comprehension. *Journal of Memory and Language* 47:292–314.

Dell, G. S. 1986. A spreading activation theory of retrieval in sentence production. *Psychological Review* 93:283–321.

Fowler, C., and J. Housum. 1987. Talkers' signaling of "new" and "old" words in speech and listeners' perception and use of the distinction. *Journal of Memory and Language* 26:489–504.

Fowler, C. A. 1988. Differential shortening of repeated content words produced in various communicative contexts. *Language and Speech* 31:307–319.

Gahl, S. 2008. *Time* and *thyme* are not homophones: The effect of lemma frequency on word durations in spontaneous speech. *Language* 84: 474–496.

Gahl, S., and S. M. Garnsey. 2004. Knowledge of grammar, knowledge of usage: Syntactic probabilities affect pronunciation variation. *Language* 80:748–775. doi:10.1353/lan.2004.0185.

Gillespie, M. 2011. Agreement computation in sentence production conceptual and temporal factors. Ph.D. dissertation, Northeastern University.

Grice, H. P. 1975. Logic and conversation. In *Syntax and Semantics*, vol.

3, *Speech Acts*, edited by P. Cole and J. Morgan, 41–58. New York: Academic Press.

Gussenhoven, C. 1983. *A Semantic Analysis of the Nuclear Tones of English*. Bloomington: Indiana University Linguistics Club.

Halliday, M.A.K. 1967. *Intonation and Grammar in British English*. The Hague: Mouton.

Heller, D., J. E. Arnold, N. Klein, and M. K. Tanenhaus. 2014. Inferring difficulty: Flexibility in the real-time processing of disfluency. *Language and Speech*. doi:10.1177/0023830914528107.

Ito, K., and S. R. Speer. 2008. Anticipatory effects of intonation: Eye movements during instructed visual search. *Journal of Memory and Language* 58:541–573. doi:10.1016/j.jml.2007.06.013.

Jurafsky, D., A. Bell, M. Gregory, and W. D. Raymond. 2001. Probabilistic relations between words: Evidence from reduction in lexical production. In *Frequency and the Emergence of Linguistic Structure*, edited by J. Bybee and P. Hopper, 229–254. Amsterdam: John Benjamins.

Kahn, J., and J. E. Arnold. 2012. A processing-centered look at the contribution of givenness to durational reduction. *Journal of Memory and Language* 67:311–325.

———. 2015. Articulatory and lexical repetition effects on durational reduction: Speaker experience vs. common ground. *Language, Cognition, and Neuroscience* 30. doi:10.1080/01690965.2013.848989.

Ladd, R. 1996. *Intonational Phonology*. Cambridge: Cambridge University Press.

Levelt, W.J.M. 1989. *Speaking: From Intention to Articulation*. Cambridge, MA: MIT Press.

Lieberman, P. 1963. Some effects of semantic and grammatical context on the production and perception of speech. *Language and Speech* 6:172–187.

Lindblom, B. 1990. Explaining variation: A sketch of the H and H theory. In *Speech Production and Speech Modeling*, edited by W. Hardcastle and A. Marchal, 403–439. Dordrecht: Kluwer Academic.

Piantadosi, S. T., H. Tily, and E. Gibson. 2011. Word lengths are optimized for efficient communication. *Proceedings of the National Academy of Sciences USA* 108:3526.

Pierrehumbert, J., and J. Hirschberg. 1990. The meaning of intonational contours in the interpretation of discourse. In *Intentions in Communication*, edited by P. R. Cohen, J. Morgan, and M. E. Pollack, 271–311. Cambridge, MA: MIT Press.

Prince, E. 1981. Toward a taxonomy of given-new information. In *Radical Pragmatics*, edited by P. Cole, 223–256. New York: Academic Press.

Schwarzschild, R. 1999. GIVENness, AvoidF and other constraints on the

placement of accent. *Natural Language Semantics* 7:141–177. doi:10
.1023/A:1008370902407.

Selkirk, E. O. 1996. Sentence prosody: Intonation, stress and phrasing. In
The Handbook of Phonological Theory, edited by J. A. Goldsmith,
138–151. Oxford: Blackwell.

Seyfarth, R. M., and D. L. Cheney. 1992. Meaning and mind in monkeys.
Scientific American 267:122–129.

Terken, J., and J. Hirschberg. 1994. Deaccentuation of words representing
"given" information: Effects of persistence of grammatical function
and surface position. *Language and Speech* 37:125–145.

Watson, D. G., J. E. Arnold, and M. K. Tanenhaus. 2008. Tic tac TOE:
Effects of predictability and importance on acoustic prominence in
language production. *Cognition* 106:1548–1557.

Watson, D. G., A. Buxó-Lugo, and D. C. Simmons. 2015. The effect of
phonological encoding on word duration: Selection takes time. In *Explicit and Implicit Prosody in Sentence Processing: Studies in Honor of
Janet Dean Fodor*, edited by L. Frazier and E. Gibson, 85–98. New
York: Springer International.

Zerkle, S., E. C. Rosa, and J. E. Arnold. 2017. Thematic role predictability
and planning affect word duration. *Laboratory Phonology: Journal of
the Association for Laboratory Phonology* 8(1): 17, 1–28, https://doi
.org/10.5334/labphon.98.

Zipf, G. K. 1929. Relative frequency as a determinant of phonetic change.
Harvard Studies in Classical Philology 40:1–95.

RELATIONAL KNOWLEDGE AND THE ORIGINS OF LANGUAGE, BENJAMIN WILSON AND CHRISTOPHER I. PETKOV

Abe, K., and D. Watanabe. 2011. Songbirds possess the spontaneous ability to discriminate syntactic rules. *Nature Neuroscience* 14:1067–74.
doi:10.1038/nn.2869 nn.2869 [pii].

Adolphs, R. 2009. The social brain: Neural basis of social knowledge.
Annual Review of Psychology 60:693.

Arnold, K., and K. Zuberbuhler. 2006. Language evolution: Semantic
combinations in primate calls. *Nature* 441:303. doi:441303a [pii]
10.1038/441303a.

———. 2008. Meaningful call combinations in a non-human primate.
Current Biology 18:R202–3. doi:S0960-9822(08)00087-0 [pii] 10.10
16/j.cub.2008.01.040.

Arriaga, G., and E. D. Jarvis. 2013. Mouse vocal communication system:
Are ultrasounds learned or innate? *Brain and Language* 124:96–116.
doi: 10.1016/j.bandl.2012.10.002.

Attaheri, A., Y. Kikuchi, A. E. Milne, B. Wilson, K. Alter, and C. I. Petkov. 2015. EEG potentials associated with artificial grammar learning in the primate brain. *Brain and Language* 148:74–80. doi:10.1016/j .bandl.2014.11.006.

Bahlmann, J., R. I. Schubotz, and A. D. Friederici. 2008. Hierarchical artificial grammar processing engages Broca's area. *Neuroimage* 42:525–534. doi:S1053-8119(08)00604-6 [pii] 10.1016/j.neuroimage.2008 .04.249.

Bahlmann, J., R. I. Schubotz, J. L. Mueller, D. Koester, and A. D. Friederici. 2009. Neural circuits of hierarchical visuo-spatial sequence processing. *Brain Research* 1298:161–170. doi:S0006-8993(09)01682-5 [pii] 10 .1016/j.brainres.2009.08.017.

Beckers, G.J.L., J. J. Bolhuis, K. Okanoya, and R. C. Berwick. 2012. Birdsong neurolinguistics: Songbird context-free grammar claim is premature. *Neuroreport* 23:139–145. doi:10.1097/Wnr.0b013e32834f1765.

Behrens, T. E., L. T. Hunt, and M. F. Rushworth. 2009. The computation of social behavior. *Science* 324:1160–1164. doi:10.1126/science.11 69694.

Belin, P., R. J. Zatorre, P. Lafaille, P. Ahad, and B. Pike. 2000. Voice-selective areas in human auditory cortex. *Nature* 403:309–312.

Bergman, T., J. C. Beehner, D. L. Cheney, and R. M. Seyfarth. 2003. Hierarchical classification by rank and kinship in baboons. *Science* 302: 1234–1236.

Berwick, R. C., G. J. Beckers, K. Okanoya, and J. J. Bolhuis. 2012. A bird's eye view of human language evolution. *Frontiers in Evolutionary Neuroscience* 4:5. doi:10.3389/fnevo.2012.00005.

Berwick, R. C., K. Okanoya, G. J. Beckers, and J. J. Bolhuis. 2011. Songs to syntax: The linguistics of birdsong. *Trends in Cognitive Sciences* 15:113–121. doi:S1364-6613(11)00003-9 [pii] 10.1016/j.tics.2011.01.002.

Bickart, K. C., C. I. Wright, R. J. Dautoff, B. C. Dickerson, and L. F. Barrett. 2011. Amygdala volume and social network size in humans. *Nature Neuroscience* 14:163.

Bickerton, D., and E. Szathmary. 2009. *Biological Foundations and Origin of Syntax*. Cambridge, MA: MIT Press.

Bornkessel-Schlesewsky, I., M. Schlesewsky, S. L. Small, and J. P. Rauschecker. 2015. Neurobiological roots of language in primate audition: Common computational properties. *Trends in Cognitive Sciences* 19:142–150. doi:10.1073/pnas.1113427109.

Chakraborty, M., S. Walloe, S. Nedergaard, E. E. Fridel, T. Dabelsteen, B. Pakkenberg, M. F. Bertelsen, G. M. Dorrestein, S. E. Brauth, S. E. Durand, and E. D. Jarvis. 2015. Core and shell song systems unique to the parrot brain. *PLoS One* 10:e0118496. doi:10.1371/journal.pone .0118496.

Cheney, D., R. Seyfarth, and B. Smuts. 1986. Social relationships and social cognition in nonhuman primates. *Science* 234:1361–1366.

Comins, J. A., and T. Q. Gentner. 2013. Perceptual categories enable pattern generalization in songbirds. *Cognition* 128:113–118.

de Vries, M. H., A. C. Barth, S. Maiworm, S. Knecht, P. Zwitserlood, and A. Floel. 2010. Electrical stimulation of Broca's area enhances implicit learning of an artificial grammar. *Journal of Cognitive Neuroscience* 22:2427–2436. doi:10.1162/jocn.2009.21385.

Dehaene, S., F. Meyniel, C. Wacongne, L. Wang, and C. Pallier. 2015. The neural representation of sequences: From transition probabilities to algebraic patterns and linguistic trees. *Neuron* 88:2–19.

Dunbar, R.I.M., and S. Shultz. 2007. Evolution in the social brain. *Science* 317:1344–1347.

Egnor, S. E., and M. D. Hauser. 2004. A paradox in the evolution of primate vocal learning. *Trends in Neurosciences* 27: 649–654. doi:S0166-2236(04)00275-9 [pii] 10.1016/j.tins.2004.08.009.

Feenders, G., M. Liedvogel, M. Rivas, M. Zapka, H. Horita, E. Hara, K. Wada, H. Mouritsen, and E. D. Jarvis. 2008. Molecular mapping of movement-associated areas in the avian brain: A motor theory for vocal learning origin. *PLoS One* 3:e1768. doi: 10.1371/journal.pone .0001768.

Fitch, W. T. 2010. *The Evolution of Language*. Cambridge: Cambridge University Press.

Fitch, W. T., and A. Friederici. 2012. Artificial grammar learning meets formal language theory: An overview. *Philosophical Transactions of the Royal Society B* 367:1933–1955.

Fitch, W. T., and M. D. Hauser. 2004. Computational constraints on syntactic processing in a nonhuman primate. *Science* 303:377–380. doi:10 .1126/science.1089401 303/5656/377 [pii].

Frey, S., S. Mackey, and M. Petrides. 2014. Cortico-cortical connections of areas 44 and 45B in the macaque monkey. *Brain and Language* 131:36–55.

Friederici, A. D. 2004. Processing local transitions versus long-distance syntactic hierarchies. *Trends in Cognitive Sciences* 8:245–247. doi:10 .1016/j.tics.2004.04.013 S1364661304001160 [pii].

———. 2011. The brain basis of language processing: From structure to function. *Physiological Reviews* 91:1357–1392. doi:91/4/1357 [pii] 10.1152/physrev.00006.2011.

Friederici, A. D., J. Bahlmann, S. Heim, R. I. Schubotz, and A. Anwander. 2006. The brain differentiates human and non-human grammars: Functional localization and structural connectivity. *Proceedings of the*

National Academy of Sciences USA 103:2458–2463. doi:0509389103 [pii] 10.1073/pnas.0509389103.

Friederici, A. D., and S. A. Kotz. 2003. The brain basis of syntactic processes: Functional imaging and lesion studies. *Neuroimage* 20, Suppl 1:S8–17. doi:S1053811903005226 [pii].

Friederici, A. D., B. Opitz, and D. Y. von Cramon. 2000. Segregating semantic and syntactic aspects of processing in the human brain: An fMRI investigation of different word types. *Cerebral Cortex* 10:698–705.

Frith, C. D. 2007. The social brain? *Philosophical Transactions of the Royal Society B* 362:671–678.

Frost, R., B. C. Armstrong, N. Siegelman, and M. H. Christiansen. 2015. Domain generality versus modality specificity: The paradox of statistical learning. *Trends in Cognitive Sciences* 19:117–125.

Gentner, T. Q., K. M. Fenn, D. Margoliash, and H. C. Nusbaum. 2006. Recursive syntactic pattern learning by songbirds. *Nature* 440:1204–1207. doi:nature04675 [pii] 10.1038/nature04675.

Gervain, J., F. Macagno, S. Cogoi, M. Peña, and J. Mehler. 2008. The neonate brain detects speech structure. *Proceedings of the National Academy of Sciences USA* 105:14222–14227.

Gómez, R. L. 2002. Variability and detection of invariant structure. *Psychological Science* 13:431–436.

Hauser, M. D., and D. Glynn. 2009. Can free-ranging rhesus monkeys (*Macaca mulatta*) extract artificially created rules comprised of natural vocalizations? *Journal of Comparative Psychology* 123:161–167. doi:10.1037/A0015584.

Hebb, D. O. 1949. *Organization of Behavior: A Neuropsychological Theory*. New York: John Wiley and Sons.

Hickok, G., and D. Poeppel. 2007. The cortical organization of speech processing. *Nature Reviews Neuroscience* 8:393–402.

Honda, E., and K. Okanoya. 1999. Acoustical and syntactical comparisons between songs of the white-backed Munia (*Lonchura striata*) and its domesticated strain, the Bengalese finch (*Lonchura striata var. domestica*). *Zoological Science* 16:319–326.

Hurford, J. R. 2007. *The Origins of Meaning: Language in the Light of Evolution*. Oxford: Oxford University Press.

———. 2012. *The Origins of Grammar: Language in the Light of Evolution II*. Oxford: Oxford University Press.

Jarvis, E. D. 2004. Learned birdsong and the neurobiology of human language. *Annals of the New York Academy of Sciences* 1016:749–777.

Jurgens, U. 2002. Neural pathways underlying vocal control. *Neurosci-*

ence and Biobehavioral Reviews 26:235–258. doi: S0149763401000689 [pii].

Kanai, Ryota, B. Bahrami, R. Roylance, and G. Rees. 2011. Online social network size is reflected in human brain structure. Paper read at Proceedings of the Royal Society B.

Lewis, P. A., R. Rezaie, R. Brown, N. Roberts, and R.I.M. Dunbar. 2011. Ventromedial prefrontal volume predicts understanding of others and social network size. *Neuroimage* 57:1624–1629.

Lu, K., and D. S. Vicario. 2014. Statistical learning of recurring sound patterns encodes auditory objects in songbird forebrain. *Proceedings of the National Academy of Sciences USA* 111:14553–14558.

Marcus, G. F., S. Vijayan, S. Bandi Rao, and P. M. Vishton. 1999. Rule learning by seven-month-old infants. *Science* 283:77–80.

Messinger, A., L. R. Squire, S. M. Zola, and T. D. Albright. 2001. Neuronal representations of stimulus associations develop in the temporal lobe during learning. *Proceedings of the National Academy of Sciences USA* 98:12239–12244.

Meyer, T., and C. R. Olson. 2011. Statistical learning of visual transitions in monkey inferotemporal cortex. *Proceedings of the National Academy of Sciences USA* 108: 19401–19406.

Milne, A. E., J. L. Mueller, C. Männel, A. Attaheri, A. D. Friederici, and C. I. Petkov. 2016. Evolutionary origins of non-adjacent sequence processing in primate brain potentials. *Scientific Reports* 6:36259.

Murphy, R. A., E. Mondragon, and V. A. Murphy. 2008. Rule learning by rats. *Science* 319:1849–1851. doi:319/5871/1849 [pii] 10.1126/science.1151564.

Neubert, F. X., R. B. Mars, A. G. Thomas, J. Sallet, and M. F. Rushworth. 2014. Comparison of human ventral frontal cortex areas for cognitive control and language with areas in monkey frontal cortex. *Neuron* 81:700–713. doi: 10.1016/j.neuron.2013.11.012.

Newport, E. L., M. D. Hauser, G. Spaepen, and R. N. Aslin. 2004. Learning at a distance II. Statistical learning of non-adjacent dependencies in a non-human primate. *Cognitive Psychology* 49:85–117. doi:10.1016/j.cogpsych.2003.12.002 S0010028504000039 [pii].

Ni, W., R. T. Constable, W. E. Mencl, K. R. Pugh, R. K. Fulbright, S. E. Shaywitz, B. Shaywitz, J. C. Gore, and D. Shankweiler. 2000. An event-related neuroimaging study distinguishing form and content in sentence processing. *Journal of Cognitive Neuroscience* 12:120–133.

Petersson, K. M., V. Folia, and P. Hagoort. 2012. What artificial grammar learning reveals about the neurobiology of syntax. *Brain and Language* 120:83–95. doi: 10.1016/j.bandl.2010.08.003.

Petersson, K. M., C. Forkstam, and M. Ingvar. 2004. Artificial syntactic violations activate Broca's region. *Cognitive Science* 28:383–407. doi: 10.1016/j.cogsci.2003.12.003.

Petkov, C. I., and E. D. Jarvis. 2012. Birds, primates, and spoken language origins: Behavioral phenotypes and neurobiological substrates. *Frontiers in Evolutionary Neuroscience* 4:12. doi:10.3389/fnevo.2012 .00012.

Petkov, C. I., C. Kayser, T. Steudel, K. Whittingstall, M. Augath, and N. K. Logothetis. 2008. A voice region in the monkey brain. *Nature Neuroscience* 11:367–374.

Petkov, C. I., Y. Kikuchi, A. E. Milne, M. Mishkin, J. P. Rauschecker, and N. K. Logothetis. 2015. Different forms of effective connectivity in primate frontotemporal pathways. *Nature Communications* 6:10. doi:10.1038/ncomms7000.

Petkov, C. I., and B. Wilson. 2012. On the pursuit of the brain network for proto-syntactic learning in non-human primates: Conceptual issues and neurobiological hypotheses. *Philosophical Transactions of the Royal Society B* 367:2077–2088. doi:10.1098/rstb.2012.0073.

Platt, M. L., R. M. Seyfarth, and D. L. Cheney. 2016. Adaptations for social cognition in the primate brain. *Philosophical Transactions of the Royal Society B* 371:20150096.

Powell, J., P. A. Lewis, N. Roberts, M. García-Fiñana, and R.I.M. Dunbar. 2012. Orbital prefrontal cortex volume predicts social network size: An imaging study of individual differences in humans. *Proceedings of the Royal Society B* 279:2157–2162.

Rauschecker, J. P. 1998. Parallel processing in the auditory cortex of primates. *Audiology and Neurootolaryngology* 3:86–103.

Ravignani, A., R.-S. Sonnweber, N. Stobbe, and W. T. Fitch. 2013. Action at a distance: Dependency sensitivity in a New World primate. *Biology Letters* 9:20130852.

Reber, A. S. 1967. Implicit learning of artificial grammars. *Journal of Verbal Learning and Verbal Behaviour* 6:855–863.

Rilling, J. K., M. F. Glasser, T. M. Preuss, X. Ma, T. Zhao, X. Hu, and T.E.J. Behrens. 2008. The evolution of the arcuate fasciculus revealed with comparative DTI. *Nature Neuroscience* 11:426–428.

Romanski, L. M., B. Tian, J. Fritz, M. Mishkin, P. S. Goldman-Rakic, and J. P. Rauschecker. 1999. Dual streams of auditory afferents target multiple domains in the primate prefrontal cortex. *Nature Neuroscience* 2:1131–1136.

Rudebeck, P. H., M. J. Buckley, M. E. Walton, and M.F.S. Rushworth. 2006. A role for the macaque anterior cingulate gyrus in social valuation. *Science* 313:1310–1312.

Rushworth, M. F., R. B. Mars, and J. Sallet. 2013. Are there specialized circuits for social cognition and are they unique to humans? *Current Opinion in Neurobiology* 23:436–442. doi:10.1016/j.conb.2012.11.013.

Saffran, J., M. D. Hauser, R. Seibel, J. Kapfhamer, F. Tsao, and F. Cushman. 2008. Grammatical pattern learning by human infants and cotton-top tamarin monkeys. *Cognition* 107:479–500. doi: http://dx.doi.org/10.1016/j.cognition.2007.10.010.

Saffran, J. R., S. D. Pollak, R. L. Seibel, and A. Shkolnik. 2007. Dog is a dog is a dog: Infant rule learning is not specific to language. *Cognition* 105:669–680. doi:S0010-0277(06)00235-6 [pii] 10.1016/j.cognition.2006.11.004.

Sallet, J., R. B. Mars, M. P. Noonan, J. L. Andersson, J. X. O'Reilly, S. Jbabdi, P. L. Croxson, M. Jenkinson, K. L. Miller, and M. F. Rushworth. 2011. Social network size affects neural circuits in macaques. *Science* 334:697–700. doi:10.1126/science.1210027.

Seyfarth, R. M., and D. L. Cheney. 1999. Production, usage, and response in nonhuman primate vocal development. In *The Design of Animal Communication*, edited by M. D. Hauser and M. Konishi, 391–417. Cambridge, MA: MIT Press.

———. 2014. The evolution of language from social cognition. *Current Opinion in Neurobiology* 28:5–9.

Seyfarth, R. M., D. L. Cheney, and P. Marler. 1980. Monkey responses to three different alarm calls: Evidence of predator classification and semantic communication. *Science* 210:801–803.

Shepherd, S. V., and M. L. Platt. 2010. Neuroethology of attention in primates. In *Primate Neuroethology*, edited by M. L. Platt and A. A. Ghazanfar, 525–549. Oxford: Oxford University Press.

Simonyan, K., and B. Horwitz. 2011. Laryngeal motor cortex and control of speech in humans. *Neuroscientist* 17 (2): 197–208. doi:10738584 10386727 [pii] 10.1177/1073858410386727.

Sonnweber, R., A. Ravignani, and W. T. Fitch. 2015. Non-adjacent visual dependency learning in chimpanzees. *Animal Cognition* 18:733–745.

Spierings, M. J., and C. ten Cate. 2016. Budgerigars and zebra finches differ in how they generalize in an artificial grammar learning experiment. *Proceedings of the National Academy of Sciences USA* 113: E3977–E3984.

Stobbe, N., G. Westphal-Fitch, U. Aust, and W. T. Fitch. 2012. Visual artificial grammar learning: Comparative research on humans, kea (*Nestor notabilis*) and pigeons (*Columba livia*). *Philosophical Transactions of the Royal Society B* 367:1995–2006.

Toro, J., and J. Trobalón. 2005. Statistical computations over a speech stream in a rodent. *Perception and Psychophysics* 67:867–875. doi:10.3758/bf03193539.

Udden, J., V. Folia, C. Forkstam, M. Ingvar, G. Fernandez, S. Overeem, G. van Elswijk, P. Hagoort, and K. M. Petersson. 2008. The inferior frontal cortex in artificial syntax processing: An rTMS study. *Brain Research* 1224:69–78. doi:10.1016/j.brainres.2008.05.070.

Udden, J., M. Ingvar, P. Hagoort, and K. M. Petersson. 2012. Implicit acquisition of grammars with crossed and nested non-adjacent dependencies: Investigating the push-down stack model. *Cognitive Science* 36:1078–1011. doi:10.1111/j.1551-6709.2012.01235.x.

Uhrig, L., S. Dehaene, and B. Jarraya. 2014. A hierarchy of responses to auditory regularities in the macaque brain. *Journal of Neuroscience* 34:1127–1132. doi: 10.1523/JNEUROSCI.3165-13.2014.

van Heijningen, C. A., J. de Visser, W. Zuidema, and C. ten Cate. 2009. Simple rules can explain discrimination of putative recursive syntactic structures by a songbird species. *Proceedings of the National Academy of Sciences USA* 106:20538–20543. doi: 0908113106 [pii] 10.1073/pnas.0908113106.

Wang, L., L. Uhrig, B. Jarraya, and S. Dehaene. 2015. Representation of numerical and sequential patterns in macaque and human brains. *Current Biology* 25:1966–1974.

Wilson, B., Y. Kikuchi, L. Sun, D. Hunter, F. Dick, K. Smith, A. Thiele, T. D. Griffiths, W. D. Marslen-Wilson, and C. I. Petkov. 2015. Auditory sequence processing engages evolutionarily conserved regions of frontal cortex in macaques and humans. *Nature Communications* 6:8901. doi:10.1038/ncomms9901.

Wilson, B., W. D. Marslen-Wilson, and C. I. Petkov. 2017. Conserved sequence processing in primate frontal cortex. *Trends in Neurosciences* 40:72–82.

Wilson, B., H. Slater, Y. Kikuchi, A. E. Milne, W. D. Marslen-Wilson, K. Smith, and C. I. Petkov. 2013. Auditory artificial grammar learning in macaque and marmoset monkeys. *Journal of Neuroscience* 33:18825–18835. doi:10.1523/JNEUROSCI.2414-13.2013.

Wilson, B., K. Smith, and C. I. Petkov. 2015. Mixed-complexity artificial grammar learning in humans and macaque monkeys: Evaluating learning strategies. *European Journal of Neuroscience* 41:568–578. doi:10.1111/ejn.12834.

PRIMATES, CEPHALOPODS, AND THE EVOLUTION OF COMMUNICATION, PETER GODFREY-SMITH

Byrne, R., U. Griebel, J. B. Wood, and J. A. Mather. 2003. Squid say it with skin: A graphic model for skin displays in Caribbean reef squid (*Sepioteuthis sepioidea*). *Berliner Paläobiol. Abh* 3:29–35.

Camp, E. 2009. A language of baboon thought? In *The Philosophy of*

Animal Minds, edited by R. Lurz, 108–127. Cambridge: Cambridge University Press.

Cheney, D. L., and R. M. Seyfarth. 1990. *How Monkeys See the World: Inside the Mind of Another Species.* Chicago: University of Chicago Press.

————. 2007. *Baboon Metaphysics: The Evolution of a Social Mind.* Chicago: University of Chicago Press.

Crawford, V. P., and J. Sobel. 1982. Strategic information transmission. *Econometrica* 50: 1431–1451.

Darmaillacq, A., L. Dickel, and J. Mather, eds. 2014. *Cephalopod Cognition.* Cambridge: Cambridge University Press.

Davidson, D. 1984. *Enquiries into Truth and Interpretation.* Oxford: Clarendon.

Dawkins, R., and J. Krebs. 1978. Animal signals: Information or manipulation? In *Behavioural Ecology: An Evolutionary Approach*, edited by J. Krebs and R. Davies, 282–309. Oxford: Blackwell.

Farrell, J., and M. Rabin. 1996. Cheap talk. *Journal of Economic Perspectives* 10:103–118.

Gallistel, R., and A. P. King. 2010. *Memory and the Computational Brain.* Chichester: Wiley-Blackwell.

Godfrey-Smith, P. 2013. Information and influence in sender-receiver models, with applications to animal behavior. In *Animal Communication Theory: Information and Influence*, edited by U. Stegmann, 377–396. Cambridge: Cambridge University Press.

————. 2014. Signs and symbolic behavior. *Biological Theory* 9:78–88.

Godfrey-Smith, P., and M. Martínez. 2013. Communication and common interest. *PLoS Computational Biology* 9:e1003282. doi:10.1371/journal.pcbi.1003282.

Grice, H. P. 1969. Utterer's meaning and intention. *Philosophical Review* 78:147–177.

Hanlon, R., and J. Messenger. 1996. *Cephalopod Behavior.* Cambridge: Cambridge University Press.

Huffard, C., R. Caldwell, and F. Boneka. 2010. Mating behavior of *Abdopus aculeatus* (d'Orbigny 1834) (Cephalopoda: Octopodidae) in the wild. *Marine Biology* 154:353– 362.

Huttegger, S., J. Bruner, and K. Zollman. 2015. The handicap principle is an artifact. *Philosophy of Science* 82:997–1009.

Lewis, D. L. 1969. *Convention: A Philosophical Study.* Cambridge, MA: Harvard University Press.

Lyon, P. 2015. The cognitive cell: Bacterial behavior reconsidered. *Frontiers in Microbiology* 6:264. doi:10.3389/fmicb.2015.00264.

Martínez, M., and P. Godfrey-Smith. 2016. Common interest and signaling games: A dynamic analysis. *Philosophy of Science* 83:371–392.

Mather, J. A. 2004. Cephalopod skin displays: From concealment to communication. In *Evolution of Communication Systems: A Comparative Approach*, edited by D. K. Oller and U. Griebel, 193–214. Cambridge, MA: MIT Press.

Mather, J. A., U. Griebel, and R. Byrne. 2010. Squid dances: An ethogram of postures and actions of *Sepioteuthis sepioidea* squid with a muscular hydrostatic system. *Marine and Freshwater Behaviour and Physiology* 43:45–61.

Maynard Smith, J., and D. Harper. 2003. *Animal Signals*. Oxford: Oxford University Press.

Millikan, R. 1984. *Language, Thought, and Other Biological Categories*. Cambridge, MA: MIT Press.

Moynihan, M., and A. Rodaniche. 1982. The behaviour and natural history of the Caribbean reef squid *Sepioteuthis sepioidea* with a consideration of social, signal and defensive patterns for difficult and dangerous environments. *Advances in Ethology* 125:1–150.

Owren, M., D. Rendall, and M. Ryan. 2010. Redefining animal signaling: Influence versus information in communication. *Biology and Philosophy* 25:755–780.

Scheel, D., P. Godfrey-Smith, and M. Lawrence. 2016. Signal use by octopuses in agonistic interactions. *Current Biology* 26:1–6.

Scott-Phillips, T. C., and R. A. Blythe. 2013. Why is combinatorial communication rare in the natural world, and why is language an exception to this trend? *Journal of the Royal Society Interface* 10:20130520. http://dx.doi.org/10.1098/rsif.2013.0520.

Scott-Phillips, T. C., J. Gurney, A. Ivens, S. P. Diggle, and R. Popat. 2014. Combinatorial communication in bacteria: Implications for the origins of linguistic generativity. *PLoS One* 9:e95929. doi:10.1371/journal.pone.0095929.

Searcy, W. A., and S. Nowicki. 2005. *The Evolution of Animal Communication: Reliability and Deception in Signaling Games*. Princeton: Princeton University Press.

Shannon, C. 1948. A mathematical theory of communication. *Bell System Technical Journal* 27:379–423.

Silk, J. B., E. Kaldor, and R. Boyd. 2000. Cheap talk when interests conflict. *Animal Behaviour* 59:423–432.

Skyrms, B. 2010. *Signals: Evolution, Learning, and Information*. New York: Oxford University Press.

Spence, M. 1973. Job market signaling. *Quarterly Journal of Economics* 87:355–374.

Sterelny, K. 2012. *The Evolved Apprentice: How Evolution Made Humans Unique.* Cambridge, MA: MIT Press.

Wagner, E. O. 2012. Deterministic chaos and the evolution of meaning. *British Journal for the Philosophy of Science* 63:547–575.

Zahavi, A. 1975. Mate selection: A selection for a handicap. *Journal of Theoretical Biology* 53:205–214.

Zollman, K., C. Bergstrom, and S. Huttegger. 2013. Between cheap and costly signals: The evolution of partial honest communication. *Proceedings of the Royal Society B* 280: 20121878. http://dx.doi.org/10.1098/rspb.2012.1878.

CONCLUSION, ROBERT M. SEYFARTH AND DOROTHY L. CHENEY

Bergman, T., J. C. Beehner, D. L. Cheney, and R. M. Seyfarth. 2003. Hierarchical classification by rank and kinship in baboons. *Science* 302:1234–1236.

Cheney, D. L., and R. M. Seyfarth. 1998. Why monkeys don't have language. In *The Tanner Lectures on Human Values*, vol. 19, edited by G. Petersen, 175–219. Salt Lake City: University of Utah Press.

———. 2007. *Baboon Metaphysics: The Evolution of a Social Mind.* Chicago: University of Chicago Press.

Cheney, D. L., R. M. Seyfarth, and J. B. Silk. 1995. The responses of female baboons (*Papio cynocephalus ursinus*) to anomalous social interactions: Evidence for causal reasoning? *Journal of Comparative Psychology* 109:134–141.

Crockford, C., R. M. Wittig, R. Mundry, and K. Zuberbuhler. 2012. Wild chimpanzees inform ignorant group members of danger. *Current Biology* 22:142–146.

Dawkins, R. 1982. *The Extended Phenotype.* Oxford: Oxford University Press.

Fisher, C., and L. R. Gleitman. 2002. Language acquisition. In *Stevens Handbook of Experimental Psychology*, vol. 3, *Learning and Motivation*, edited by H. F. Pashler and C. R. Gallistel, 445–496. New York: Wiley.

Fitch, W. T. 2010. *The Evolution of Language.* Cambridge: Cambridge University Press.

Gautier, J. P., and A. Gautier. 1977. Communication in Old World monkeys. In *How Animals Communicate*, edited by T. Sebeok, 890–964. Bloomington: Indiana University Press.

Marler, P. 1983. Monkey calls: How are they perceived and what do they mean? In *Advances in the Study of Mammalian Behavior*, edited by

J. F. Eisenberg and D. G. Kleiman, 7:343–356. Stillwater, OK: American Society of Mammalogists [Special Publication Series].

Schel, A. M., S. W. Townsend, Z. Machanda, K. Zuberbuhler, and K. E. Slocombe. 2013. Chimpanzee alarm call production meets key criteria for intentionality. *PLoS One* 8:e76674.

Schlenker, P., E. Chemla, A. M. Schel, J. Fuller, J.-P. Gautier, J. Kuhn, D. Veselinovic, K.Arnold, C. Casar, S. Keenan, A. Lemasson, K. Ouattara, R. Ryder, and K. Zuberbuhler. 2016. Formal monkey linguistics. *Theoretical Linguistics* 42:1–90.

Seyfarth, R. M., and D. L. Cheney. 2014a. The evolution of language from social cognition. *Current Opinion in Neurobiology* 28:5–9.

———. 2014b. The evolution of concepts about agents: Or, what do animals recognize when they recognize an individual? In *Concepts: New Directions*, edited by E. Margolis, 57–76. Cambridge, MA: MIT Press.

———. 2016a. Precursors to language: Social cognition and pragmatic inference in primates. *Psychonomic Bulletin and Review*. doi:10.3758/s13423-016-1059-9.

Silk, J. B., S. C. Alberts, and J. Altmann. 2003. Social bonds of female baboons enhance infant survival. *Science* 302:1231–1234.

Silk, J. B., J. C. Beehner, T. Bergman, C. Crockford, A. L. Engh, L. Moscovice, R. M. Wittig, R. M. Seyfarth, and D. L. Cheney. 2009. The benefits of social capital: Close social bonds among female baboons enhance offspring survival. *Proceedings of the Royal Society B* 276: 3099–3104.

———. 2010a. Female chacma baboons form strong, equitable, and enduring social bonds. *Behavioral Ecology and Sociobiology* 64:1733–1747.

———. 2010b. Strong and consistent social bonds enhance the longevity of female baboons. *Current Biology* 20:1359–1361.

Spelke, E. S., and K. D. Kinzler. 2007. Core knowledge. *Developmental Science* 10: 89–96.

Stevens, J. S., L. R. Gleitman, J. Trueswell, and C. Yang. 2017. The pursuit of word meanings. *Cognitive Science* 41:638–676.

Zuberbuhler, K., R. Noe, and R. M. Seyfarth. 1997. Diana monkey long distance calls: Messages for conspecifics and predators. *Animal Behaviour* 53:589–604.

INDEX

acoustic prominence: *vs.* acoustic reduction, 63–65; production, 65–68; speech production fluency, 68–74

acoustic reduction, 72; acoustic prominence *vs.*, 63–65

Amboseli National Park (Kenya), 3

American Revolution, 109

American Sign Language, 57

amygdala, 98

animal: Artificial Grammar Learning paradigm and animal behavior, 87–93; combinatorial structure in sign systems, 103–4; communication of, 2, 4; evolution of communication, 102; learning relationships between events, 82–83; reproductive success in, 29–30; responses to legal and illegal sequences, 88–94

animal vocalizations, 10–11: communication during cooperative interactions, 16–18; social function of, 15–30; theoretical background, 15–16

ape language, 12

Arnold, Jennifer E., 4, 62–78

Artificial Grammar (AG), 5; animal sequence processing, 87–94

Artificial Grammar Learning (AGL) paradigms, 81–83; animal behavior and, 87–92, 100; language processing, 95–98; sequence processing, 81–83, 92–94; social knowledge, 92–94

audience, 73, 133n3

auditory receptive learning, 80, 94

Australian giant cuttlefish (*Sepia apama*), 116

avian alarm, 127

Baboon Metaphysics (Seyfarth and Cheney), 3, 47, 115, 117

baboons, 3, 5–6; cognition continuities, 26–30; cognitive abilities, 47–50, 126; combinatorial system of communication, 128–29; communication, 103, 120; dominance hierarchy, 84–86, 99; field studies of social knowledge and, 92–94; language evolution, 47–50; listener recognizing caller, 20–21; sequence processing, 92–94; signaling, 114; social knowledge, 83, 92–94; social rank and relational knowledge, 84–87; vocalizations, 15, 17, 18–30, 85–87, 103, 125

Botswana (Okavango Delta), 3, 18, 83, 92

brain: language processing, 95–99; mechanisms in humans and nonhuman primates, 11–13; size, 25, 100

Broca's territory, 95–96

camouflage, 116, 119

Camp, Elisabeth, 115

Campbell monkeys, 132n6

Cantonese sentence, 43–44

cephalopods, 5–6; camouflage, 119; signal complexity of, 117–20; skin patterning, 103–4, 115–16

Cercopithecus diana (Diana monkeys), 127

Cheney, Dorothy L., 3, 9–33, 47, 102, 103, 113, 115, 123–29

Chimpsky, Nim, 59–60

Chomsky, Noam, 1, 4, 37, 40, 48, 49, 51, 131nn1, 2, and3; Chomskyan syntactocentrism, 4, 41

chromatophores, 116, 119

Clark, Herbert, 13–15

coevolution: framework, 120; sender-receiver, 104–9, 111, 115; sign production and interpretation, 102

coevolve: rule of production, 114; sending rule, 108, 111; term, 102

common interest: sender-receiver model, 105–7

communication: act-to-act coordination, 105; act-to-state coordination, 105; animal vocalizations, 15–30; brain mechanisms, 11–13; cognition continuities, 26–30; combinatorial system of, 128–29; during cooperative interactions, 16–18; differences between human and nonhuman primate, 9–10, 24–26, 32–33, 46, 62–64, 79–80; evolution by natural selection, 102–3; hierarchical syntax, 131–32n4; importance of pragmatics, 124–26; organized sign systems and combinatorial structure, 109–13; revealing knowledge of baboons, 22–23; sender-receiver coevolution, 104–9

communicator, 103, 133n1, 133n3

Connors, Jimmy, 67

continuity: constructions of human language, 46–47; of grammar, 57–61; small and specific hypotheses, 50–57

creole languages, 4, 38, 42–44

crowned eagles (*Stephanoaetus coronatus*), 127

cue: definition, 108, sender-receiver models, 107–9

Darwin, Charles, 55, 56

deviance detection (oddball) paradigm, 97

Diana monkeys (*Cercopithecus diana*), 127

disfluency, 67–68, 71, 73, 133n1

Dobzhansky, Theodosius, 1

Duke Institute for Brain Sciences, 2

empathy, 2, 38

English, 39, 51: prosody of, 75; verb-noun compounds of, 52–54

evolution: of call production, 114; of language, 1, 47–50, 51–52, 60–61, 79, 132n9; by natural selection, 102–3; organized sign systems and combinatorial structure, 109–13

exploitation, 108, 133n7

face recognition, 11

face-to-face interaction, 23–24

fluency, 60; human language, 62–78; information status and, 74–78; prosody, 72

forest monkey, 127–28, 132n6

functional Magnetic Resonance Imaging (fMRI), 95, 97

Genesis 1:1, 1

Godfrey-Smith, Peter, 5–6, 102–20

grammar, 3, 75; continuity of, 57–61; intentional influence, 75–76; morphology of, 40; pragmatic usage of, 37–39; protostages of, 51; verb-noun compounds, 52–55

grammaticalization, 38–39, 50, 51, 132n9

grammatical relationships, 80, 96, 129

grammatical rules, 9

Hebbian learning, 82–83

honest signaling hypothesis, 16

How Monkeys See the World (Seyfarth and Cheney), 3

human language: continuity in, 46–47; evolution, 1, 47–50, 51–52, 60–61, 79; fluency and information status, 74–78; fluency effects, 62–78; pathways for future study, 99–101; proposals, 1–2; rule-governed system, 62; sequence ordering, 97; speech fluency and acoustic prominence, 68–74; syntax, 80–81; word duration, 70

humans: brain and social cognition, 98–99; brain mechanisms, 11–13

human syntax: recursive structures, 50; VP/SC (verb phrase/small clause), 52. *See also* syntax

information status: acoustic prominence production, 65–68; acoustic prominence *vs.* acoustic reduction, 63–65; fluency and, 74–78

lactose tolerance, 132n8

language: Cantonese, 43–44; difference between human and animal communication, 126–27; neurobiological substrates and processing, 94–99; pragmatics and, 39–40; Russian, 42–43; Saramaccan creole, 44; semantic parity, 126–29; social cognition as precursor of, 30–32; social function of, 13–15; study of use, 14–15. *See also* human language

language mutation, 131n1

language of thought, 48, 115, 129

lanterns: signals by sexton, 109–11

learning: vocal production and auditory receptive, 80

Lewis, David, 104, 133n3; pattern, 104–5

linguistics, 2, 4; grammar, 37–39; importance of pragmatics, 124–26; languages and pragmatics, 39–40; modern developments and pragmatics, 41–45

living fossils, 4, 52

Macaca mulatta (rhesus macaque), 17

McWhorter, John, 4, 37–45

Martínez, Manolo, 106, 120

Mather, Jennifer, 118

Minimalism, 51

mirror neurons, 12

morphophonemics, 37, 43

Moynihan, Martin, 117–19, 134n14

Nash equilibrium, 107

natural: term, 134n9

natural relation: term, 109

natural selection: animal vocalizations, 15–16; evolution by, 102–3; favoring cognition, 26–30

neurobiology, 2, 79; language processing, 94–99; pathways for future study, 99–101

New World monkeys, 90, 91

nonhuman primates: bonobo Kanzi, 57–58, 132n7; brain mechanisms, 11–13; chimpanzee Washoe, 57–58; communication, 9–10; gorilla Koko, 55; sequence processing, 87–92; social cognition, 98–99; social function of vocalizations, 18–30; vocalizations, 10–11

nonsyntactic stage, one-word, 58

objects: human cognition, 12–13
octopuses, 116; signaling, 116
Okavango Delta (Botswana), 3, 18, 83, 92
Old World monkeys, 13, 90, 91, 94
orbitofrontal cortex, 98
ostensive communication, 2, 41

Papio cynocephalus ursinus (wild baboons), 18
Peirce, C. S., 133n3
Petkov, Christopher I., 5, 79–101
philosophy, 2, 4, 104, 120
phonemes, 9, 63
pidgin languages, 4, 40, 42, 44
pitch, 4, 63, 64, 75–77, 88, 97
Planer, Ron, 120, 134n9
Platt, Michael L., 1–6
Poecile atricapillus (black-capped chickadee), 26
pragmatics: importance of, 124–26; languages and, 39–40; modern developments and, 41–45
private information, 105
Progovac, Ljiljana, 4, 46–61
pronunciation: speech difficulty, 71–72; variability in, 63
prosody, 4, 67; duration, 69, 76; functional use of, 76; grammatical constraints of, 78; information status, 75; language dimension, 63; pitch accent, 75; speech difficulty, 71–72; timing component of, 69
protolexicon, 51
protostages, grammar, 51
protosyntactic (two-word) stage, 52–53, 58
protosyntax, 4, 132n5
psychology, 2, 4, 111

reef squid (*Sepioteuthis sepioidea*), 117–19

relational knowledge, baboon social rank and, 84–87
Revere, Paul, 109
Rodaniche, Arcadio, 117–19, 134n14
Russian language, 40, 42–43

Saramaccan language, 44
Scott-Phillips, Thom, 112, 118
screams, 19, 22, 24, 26–28, 85, 113, 128
semantics, 1, 3; parity, 126–29; pragmatics and, 4, 6, 41; representation 13; sequencing structure without, 82; syntax and, 1, 6, 124, 126
sender-receiver coevolution: combinatorial structured sign, 112–13; common interest between sender and receiver, 105–7; communication, 104–9; Lewis pattern, 104–5; sending rule, 108, 111–12; sexton's use of lanterns, 109–11; signals and cues, 107–9; "zero sum" relation, 133n5
sender-receiver models, 107–9
sentence, 1, 9, 71, 73; Cantonese, 43–44; construction of, 45, 131n3; grammatical relationships, 80–82; inferring speaker's meaning, 62–64, 66, 68; meaning, 128–29; pronunciation of, 76–77; reconstruction of, 51; Russian, 40, 42–43; Serbian, 52–53; syntactic framework, 131n3
Sepia apama (Australian giant cuttlefish), 116
Sepioteuthis sepioidea (reef squid), 117–19
Serbian language, 52–53
sexton: lanterns and signals, 109–11
Seyfarth, Robert, 2–3, 9–33, 47, 102, 103, 113, 115, 123–29

signal, 109; definition of, 108; sender-receiver models, 107–9
signaling: coleoids, 116–17; complexity of cephalopods, 117–20
signal transduction, 109
sign systems, 109–13, 127, 134n10
skin patterning, 134n14; cephalopods, 103–4, 115–16
social cognition: human language, 37; precursor of language, 30–32
social ecology, 6, 114
social function: animal vocalizations, 15–30; continuities in, 23–26; of language, 13–15
songbirds, 79, 91, 94, 96, 99
speech, 12: acoustic properties of, 4–5, 63; disfluency of, 67–68; fluency and acoustic prominence, 68–74; running, 65; thought as inner, 48
Spence, Michael, 107
Stephanoaetus coronatus (crowned eagles), 127
swear words, 56–57
syntactocentrism, 4, 41
syntax, 1, 4–6, 110, 117; communication as hierarchical, 131–32n4; evolution of, 52, 54; as grammar, 37–38, 46; grammatical relationships between words, 80–81; phonology or, 38, 40; pragmatics and, 41, 44, 124

Tac Tac Toe, 71
temporal lobe, 96, 98
theory of mind, 2, 25
threat-grunt, 19–22, 24, 26–28, 85, 113–14

ventromedial prefrontal cortex, 98
verb-noun compounds, 132n5; grammar, 52–55
vocalizations: animals, 10–11; call meaning, 21–22; continuities in cognition, 26–30; listener recognizing caller, 20–21; nonhuman primate, 18–30, 85–87; screams, 19, 22, 24, 26–28, 85, 113, 128; social function continuities, 23–26; swearwords, 56–57; threat-grunt, 19–22, 24, 26–28, 85, 113–14
vocal production: of baboons or macaques, 94; of humans and animals, 79–80; learning, 80, 99–100; timing of, 10

Wilson, Benjamin, 79–101

Zahavi, Amotz, 107